Globalization, Violence and Security

Local Impacts of Regional Integration

P.I.E. Peter Lang

Bruxelles · Bern · Berlin · Frankfurt am Main · New York · Oxford · Wien

Shirlita Espinosa & Antonella Fazio (eds.)

Globalization, Violence and Security

Local Impacts of Regional Integration

Regional Integration and Social Cohesion
No.16

With the support of the Consortium for Comparative Research on Regional Integration and Social Cohesion (RISC).

The book was subject to a double blind refereeing process.

Éditions scientifiques internationales
Brussels, 2016
1 avenue Maurice, B-1050 Bruxelles, Belgique
www.peterlang.com ; info@peterlang.com

Printed in Germany

ISSN 2030-8787
ISBN 978-2-87574-355-8
eISBN 978-3-0352-6629-0
D/2016/5678/49

CIP available from the British Library and from the Library of Congress, USA.

Bibliographic information published by "Die Deutsche Nationalbibliothek"
"Die Deutsche Nationalbibliothek" lists this publication in the "Deutsche Nationalbibliografie"; detailed bibliographic data is available on the Internet at http://dnb.d-nb.de.

Table of Contents

Preface

Harlan Koff

On 8 January 2016, Joaquín Archivaldo Guzmán Loera, known universally as "El Chapo," was arrested by Mexican authorities during a raid in the Mexican state of Sinaloa. The arrest of one of the world's most recognized leaders of organized crime was celebrated by state officials in Mexico and the United States, amongst others as an important victory in the "war on drugs." While the symbolism of this arrest is significant, it is important to maintain perspective. Enrique Peña Nieto, the President of Mexico, has received some criticism amongst the applause for the recapture of El Chapo for celebrating the importance of this arrest too prominently. In an article for CNN, award-winning author Don Winslow wrote:

> Anytime a major "kingpin" is captured, we see it as a victory in the war on drugs. We could decorate a long wall with posters of former kingpins: Miguel Angel Gallardo, Carlos Lehder, Griselda Blanco, Frank Lucas, Benjamin Arellano-Felix, Pablo Escobar of recent "Narcos" fame, Amado Carrillo Fuentes, the former "Lord of the Skies", Osiel Cardenas and now Joaquin Guzman. I indulge in a list because it's instructive. Each of these captures was supposed to bring victory in the war on drugs and each of these victories has resulted in absolutely nothing. (<http://edition.cnn.com/2016/01/08/opinions/winslow-el-chapo-capture/>).

In fact, if historical trends hold true, then the newest arrest of El Chapo may actually increase violence in Mexico rather than diminish it in the same way that the killings of Pablo Escobar in 1993 and Osama bin-Laden in 2011 led to power vacuums that actually increased violence related to drug trafficking in Colombia and Islamic terrorism.

Winslow's statement, cited above, reminds us that the "war on drugs" is a never-ending conflict. Recent scholarship on this subject has informed this position in detail. For example, in the RISC Consortium's 2008 volume, *Perspectivas Comparativas sobre el Liderazgo*, Maria de Lourdes Dieck Assad and Jacobo Ramirez (2009) criticized then-Mexican President Felipe Calderon's war on drug trafficking for attempting to address this problem through military tactics. Their chapter contended that Mexico's drug war is linked to social cohesion problems in the country and that only policies that focus on these underlying causes of

Mexico's recent drug-related violence could effectively improve security issues in the country.

This volume focuses on violence and security in a similar way. While discussions centered on organized crime, trafficking and terrorism grab headlines, they also limit debates on these subjects. The world has witnessed a securitization of politics since the September 11, 2001 attacks on the World Trade Center and the Pentagon. Terrorist attacks in Europe, most recently in Paris on 13 November 2015 have contributed to the global reach of the securitization of politics. The year 2016 has further extended attention to terrorism as attacks have occurred in Jakarta, Indonesia and Ouagadougou, Burkina Faso.

We must ask, however, whether we are not oversimplifying policy responses to insecurity and violence. National leaders in many parts of the world, such as Mexico, have declared wars on drugs and terror. Moreover, debates over migration in most advanced industrial states have been infused with protectionist discourses related to security. Even emerging states, such as South Africa, have reacted defensively in the face of immigration and such nativist positions have contributed to high-profile episodes of violence. Like Dieck Assad and Ramirez, numerous authors have indicated that the securitization of politics has led to violent responses from organized crime and terrorism and increased human vulnerability (see Dunn, 1996). Rather than pursuing security by reinforcing social cohesion in places characterized by violence, contemporary policy-responses usually focus specifically on violence and often this means the securitization of non-security policy arenas, such as development (see Gabrielli, 2007), migration (see Cornelius *et al.*, 2004), border policies, (see Andreas, 2000) etc. This scholarship reminds us that violence and security are not synonyms but separate conceptual paradigms and political realities. In short, insecurity breeds violence and by focusing specifically on the latter, policy-makers are oversimplifying security issues.

Our political focus on violence has also led to the reinforcement of what Michel Foucault (1986) calls "areas of fracture," defined as policies that highlight power differences between developed and developing states. These fractures are very important for our understanding of violence. Popular images of violence generally associate it with underdevelopment and poverty. However, recent trends indicate that this is not necessarily true. Terrorism, organized crime and trafficking are multibillion dollar industries where participant organizations resemble multinational corporations more than gangs. Emerging economies often foster organized crime because of the presence of a dangerous mix of weak state institutions, political parties and leaders that lack legitimacy, the establishment of international distribution channels, cutting edge technology in illicit criminal markets, and most importantly, increasingly prominent social divides as development often leads to the concentration of capital in particular sub-regions or amongst specific groups (see Weisbront, Lefebvre and Sammult, 2014). Countries such as Mexico, South Africa, Colombia, Turkey, Indonesia, etc. are witnessing unprecedented levels of economic development combined with more prominent social cleavages. For example, South America is a region witnessing elevated economic growth rates combined with increasing social divisions (as evidenced by the highest Gini coefficient scores in the world) and rising homicide rates (see Ascher and Mirovitskaya, 2012). For this reason, the current literature on violence focuses heavily on these countries "in transition" (see for example, Mickunas, 2016) and this scholarship suggests that insecurity and violence seem to be characteristics of contemporary development models more than symptoms of underdevelopment. This point is a common thread that links the articles included in this volume.

This observation is also highly relevant for our understanding of development in contemporary affairs because the world has witnessed increasing securitization of development cooperation despite universal agreement on the Sustainable Development Goals (SDGs) in 2015, which focus on "transformative development" aimed at promoting equity in the world. One problem related to these "international objectives" is that they are undermined by "transnational policies." To begin with, the regionalization of development cooperation has securitized non-security policy arenas across national borders. This has occurred within regions as organizations such as the European Union (EU), the Economic Community of West African States (ECOWAS) and the Association of Southeast Asian Nations (ASEAN) have developed further-reaching regional security strategies.

More importantly, important development donors, such as the European Union and the United States (US) have securitized development aid through the introduction of conditionality in aid and trade agreements. Scholarship on the Union for Mediterranean (see Carrera, Den Hertog, and Parkin, 2012), EU development aid to African regions and states (see Fisher and Anderson, 2015) and US development aid programs in Mexico and Central America (see Sandoval, 2006) have generally indicated that development aid has been utilized to promote foreign policy objectives related to security rather than vice versa.

The RISC Consortium has critically analyzed the relationships between social cohesion, development and security since its establishment in 2007 because it recognizes the intrinsic relationship between violence and economic development models that do not account for social cohesion.

Growth without cohesion leads to social divisions that inevitably impact security, which leads to increased violence. RISC's working group on "Conflict, Violence and Citizen Security" based in the Instituto de Estudios Políticos of the Universidad de Antioquia in Colombia has organized various events and published empirical research on the importance of social cohesion to the success of security strategies. RISC has also organized a conference for the Latin Americanists of Belgium and Luxembourg (LABEL) on "Human and Environmental Security in Cross-Border Regions: Multidisciplinary Approaches in Latin America" which highlighted the need to understand socio-economic transformations that globalization has fostered in cross-border regions in Latin America. These developments, such as the emergence of the maquiladora industry along the US-Mexico border, have contributed to contexts of violence, such as the femicide which has been occurring in and around Ciudad Juarez, Mexico.

In fact, this research agenda indicates that broader security policies are better placed to address violence than the narrow "wars" (on drugs, terror, trafficking, etc.) that seem to guide much policy. In 2014, RISC organized different events on human and environmental security in cross-regional perspective. These seminars and publications highlighted the importance of implementing transnational human and environmental security strategies through regional development aimed at reinforcing social cohesion and human rights. This approach contends that the militarization of development and border policies actually exacerbate human insecurity and violence by contributing to the proliferation of social cleavages (see Koff and Maganda, 2016).

This volume is an important contribution to RISC's ongoing agenda on violence and security issues and their relationship to development and social cohesion. It results from the consortium's 2012 International Conference on "Globalization, Violence and Security: Local Impacts of Regional Integration" which was organized by the Centro de Investigaciones y Estudios Superiores en Antropología Social (CIESAS) and the Technólogico de Monterrey (ITESM) in Mexico City. This conference examined the decisive impacts and proposed local, regional and global solutions to insecurity and violence in comparative perspective. The starting point was a risk agenda impacting countries and regions which challenges capabilities to integrate social answers and public agendas responding to critical issues such as trafficking in drugs, humans, weapons, organs and money, as well as decisive factors such as inequality, exclusion, and a lack of opportunities. Some of the features that contemporary societies share are: a) inability or weakening of the State to keep control over the territory as well as the monopoly of the legitimate use of force; b) penetration or cooptation of State institutions by criminal organizations, c) the fragility of citizens fearful of both assaults by criminal groups and by the inability of the State to guarantee their rights, d) growing social stratification and segmentation in the illicit drug "consumption" countries, in addition to scarce citizen participation in response to public health problems and criminality, e) the lack of responsibility taken by countries and regions characterized by consumption, of drugs, organs, and prostitution and the sale of weapons for the systemic impacts of their actions, and f) lack of standards, institutions as well as regional and global public policies with a comprehensive perspective for reducing violence and insecurity. RISC thanks Dr. Bernardo González-Aréchiga and Dr. Elena Azaola Garrido who were the main scientific organizers and intellectual promoters of the event.

The following chapters build on the aforementioned themes. The first two contributions, by Vanessa Barolsky, and Claudia Puerta Silva and Alejandro Agudelo Calle address the speed of violence through political and cultural transmission. Barolsky presents an analysis of the discursive production of knowledge on violence. Her research investigates published reports on violence by international agencies that define what violence is and how much violence there is globally. Barolsky rightly questions the legitimacy of these reports and how a scientific study of violence is very much influenced by regimes of knowledge surrounding violence. She particularly points out that the anxiety to relentlessly secure the developed world has burdened the "violent" global South. This is followed by Puerta Silva and Agudelo Calle's chapter on the intersection between Colombia's television shows as civic education and the state of the public health system. The chapter reflects on the results of a study in 2012 that involves an analysis of four television programs and semi-structured interviews with viewers of these programs. The authors argue that while television programs have the potential to educate citizens, they are focused on the disease and its prevention and do not sufficiently address Colombia's health system and its failures which the authors see as a form of structural violence that hurt Colombians.

This section is followed by two chapters that specifically discuss violence in urban contexts in emerging states. Carlos Mario Perea Restrepo examines the relationship between violent conflict, the city and the peace process between the Colombian government and FARC that began in 2012. He argues that the agenda of the peace process does not include any reference to urban violence. This omission is very telling of what could potentially fail in the peace process in cities where three-quarters of Colombians. These urban places have witnessed intense and prolonged violence. While the rural-focused peace process agreement is understandable due to the weight of historical factors, the author posits that the necessity to include the city must not be overlooked. Addressing inequalities and violence in the city may greatly deepen the positive social changes for which the peace process is aiming.

Joana Vargas and Natasha Elbas Neri then present a chapter on extra-judicial punitive practices of young adults in Rio de Janeiro in Brazil. The research presents statistics and interviews with young offenders to show the intense violence confronting young people of the *favelas* in Brazil. The authors argue that police violence in Brazilian *favelas* harm mostly black and mixed-race male youths and that it is a result of social inequalities and the lack of the political will of the state to solve the problem. Furthermore, the elite political class uses the fear of crime and urban violence to gain leverage.

Finally, part three of the book presents two chapters that link violence with identity issues and development. Robert Dover, Paula Andrea Hinestroza and Gloria Patricia Lopera Mesa present a study of development and violence related to regional and multinational mining identities in Colombia. Presented here are the findings from an 18-month project in Marmato where large scale resource extraction and infrastructure development threaten to affect indigenous communities. In particular the illegalization of traditional mining threatens to undermine of the way of life for these populations. The authors argue that events in Marmato created a transformation in local identities as well, which may either be adaptation to the rules of the new actors or resistance against them. Finally, the collection presents Emeline Nanga's chapter discussing the different types of violence that migrants endure in Mexico and Morocco within the context of regional integration. Nanga's work focuses on the political pressures that emerging states, such as Morocco and Mexico receive to securitize migration, which leads to increased violence against migrants in these new receiving/transit states.

These studies present differing perspectives on violence, its meaning, transmission mechanisms and its relationship to development. Together, they present a comparative view of this salient issue in contemporary global affairs.

References

Andreas, P. (2000). *Border Games: Policing the U.S.-Mexico Divide.* Ithaca: Cornell University Press.

Ascher, W. and Mirovitskaya, N. (2012). *Economic Development Strategies and the Evolution of Violence in Latin America.* London: Palgrave Macmillan.

Carrera, S., Den Hertog, L. and Parkin, J. (2012). 'EU Migration Policy in the Wake of the Arab Spring, What Prospects of EU-Southern Mediterranean Relations?' *MEDPRO Technical Report* No. 15. Brussels: MEDPRO.

Cornelius, W., Tsuda, T., Hollifield, J. and Martin, P. (eds.) (2004). *Controlling Immigration.* Palo Alto, CA: Stanford University Press.

Dieck Assad, M. and Ramirez, J. (2009). 'Violence and Ungovernability in Latin America and the Caribbean. The Mexican Case.' In C. Maganda and H. Koff (eds.) *Perspectivas comparativas del Liderazgo*. Brussels: P.I.E. Peter Lang.

Dunn, T. J. (1996). *The Militarization of the U.S.-Mexico Border, 1978-1992: Low Intensity Doctrine Comes Home*. Austin: University of Texas Press.

Fisher, J. and D. Anderson (2015). 'Authoritarianism and the Securitization of Development in Africa' *International Affairs* 91 (1): 131-151.

Foucher, M. (1986). *L'Invention des Frontières*. Paris: La Documentation Française.

Koff, H. and Maganda, C. (2016). 'Environmental Security in Transnational Contexts: What Relevance for Regional Human Security Regimes?' *Globalizations* 13 (6): 1-11.

Mickunas, A. (2016). 'Globalization, Neoliberalism, and the Spread of Economic Violence: The Framework of Civilizational Analysis.' In V. Berdaves and J. Murphy (eds.) *Neoliberalism, Economic Radicalism, and the Normalization of Violence*. Cham Switzerland: Springer.

Sandoval, J.-M. (2006). 'Migración y seguridad nacional en las fronteras sur y norte de México', In D. Villafuerte Solis y X. Leyva Solano (eds.) *Geo-economía y Geopolítica en el área del Plan Puebla Panamá*, Mexico: Camera de Deputados, LVII Legislatura.

Weisbrot, M., Lefebvre, S. and Sammut J. (2014). 'Did NAFA help Mexico? An Assessment after 20 Years.' *Center for Economic and Policy Research*. <http://www.cepr.net/documents/nafta-20-years-2014-02.pdf>. [Accessed 20 January 2016].

Part 1

Violence and the Production of knowledge

Violence: Global Solutions to a Global Problem?

Vanessa BAROLSKY

Introduction

This paper investigates the growing number of reports on violence produced by a variety of global multi-lateral institutions over the last decade. The paper argues that this increase in global reports on violence is not simply a reflection of an empirical rise in violence but is instead the result of systemic processes of knowledge production that are deeply implicated with power and "produce the reality they seem to reflect" (Stern and Öjendal, 2010: 7). The paper therefore asks whether these "global" understandings and "global" responses to violence are in fact "global" as in universal or do they reflect the understandings and aspirations of a small number of powerful global actors? Hence this paper seeks to interrogate the discursive production of knowledge about violence in reports produced by global multi-lateral institutions that configure the world into zones of risk and disorder and zones of modernity and rationality, through claims to universal knowledge and "scientific" neutrality. The paper instead argues that the specificity of conditions in the global south[1] in fact contest many of the assumptions embedded in this global knowledge production. At stake in these global constructions of knowledge are enormously significant funding flows that have the power to shape national states' responses to violence and, which can also become the premise of international interventions that sometimes exacerbate, rather than alleviate violent conflict in arenas around the world.

A new insecurity

In order to understand how knowledge about violence is produced, this paper seeks to systematically investigate the meta-level of global knowledge production about violence and security at a local level. While

1 There is much debate about the "global south" as an analytic construct and object of investigation. Here it is used in the loose relational sense suggested by Comaroff and Comaroff (2012) as a term that refers to an unstable and porous "ex-centric" location to "hegemonic centres" (p. 115).

globalisation has brought the "local" and "global" into irrevocable interaction with each other, the nature of these connections, the nature of modernity this intersection is creating, and the type of future it augurs, remain deeply contested. Although there is literature which interrogates the discursive construction of a relationship between security and development in many global multi-lateral institutions, there is very little literature which critically investigates the normative assumptions about state and society within policy discourses on violence in relation to the social practices and ontological orientations that actually exist in societies that are the object of global interventions and "corrections". These may "deviate" in important ways from the Western ideal of the autonomous, rational, calculating self and a Westphalian model of the statehood in which all forms of social regulation are formalised, neutralised and bureaucratised. Thus far the normative vision of modernity that shapes global discourses on violence and the forms of intervention that these discourses propose, have seldom been openly and critically interrogated by those working in the field of violence prevention as this knowledge production is often deeply integrated with funding flows. This paper attempts to begin to open up this discussion so that the voices of the people whose lives are so often written about can meaningfully shape how we understand "what works" and what doesn't "work".

The paper focuses on the forms of knowledge that are articulated and shaped in the wide array of global institutions that seek to address the challenge of violence, including organisations such as the World Health Organisation, the World Bank, the Geneva Declaration Secretariat, UN Habitat, United Nations Development Programme (UNDP), the Organisation for Economic Co-operation and Development (OECD), the European Union and other organisations.

The first decade of the millennium appears to have given rise to a multiplicity of renewed international efforts to understand, interpret and shape responses to the forms of the violence and "insecurity" which have characterized the era of globalisation. Comaroff and Comaroff, for example, noted in 2004 an "uncommon preoccupation with public order, crime, and policing" (Comaroff and Comaroff, 2004: 800). Just some examples of this global concern include the 2002 World Health Organisation report, *World Report on Violence and Health*; the UNICEF *World Report on Violence against Children* in 2006; the *Global Burden of Armed Violence Reports* in 2008 and 2011; the World Bank *World Development Report: Conflict, Security and Development 2011*, the UNODOC report, *Global Study on Homicide Trends 2011*, a 2013 WHO report on violence against women, *Global and Regional Estimates of Violence Against Women* (2013) and a 2014 report by UNICEF on violence against children, *Hidden in Plain Sight*.

It is argued here that this knowledge production cannot simply be seen as responding to a sudden empirical rise in global levels of violence. However, what is evidenced by these reports is a sense that violence in the new millennium presents a different type of challenge to the violence of the 20th century and that there is an urgent need to understand these types of violence. As Anthony Giddens has argued, "Our age is not more dangerous – not more risky – than those of earlier generations – but the balance of risks and dangers has shifted" (Giddens, 2000: 52). Thus the World Development report notes the complexity of violence in the 21st-century, which does not fall into "neat categorisations of 'war' or 'peace', or into 'criminal violence' or 'political violence'" (The World Bank, 2011: 3). In this context, "New threats – organised crime and trafficking, civil unrest due to global economic shocks, terrorism – have supplemented continued preoccupations with conventional war between and within countries" (The World Bank, 2011: 1).

This paper investigates the claims of the institutions that produce global reports on violence to know "what works" through an analysis of the discourses in some of the most globally influential reports produced in the past decade about violence. These include reports by WHO, the Geneva Declaration Secretariat, the World Bank and UNODOC. The World Health Organisation's interpretation of violence as a public health problem and the definition of violence outlined in its 2002 World Report, have shaped the analysis and response, particularly to interpersonal violence, of countless interventions around the world since the publication of the report. The Geneva Secretariat *Global Burden of Armed Violence* report of 2008 was the seminal study linking the problems of violence and development, which subsequently became a standard assumption in much development discourse. The World Bank's *World Development Report* of 2011 provided the most systemic conceptual engagement with the link between conflict, security and development in the policy area thus far and has continued to be extremely influential in shaping thinking on these links. The report by UNODOC on homicide trends, also produced in 2011, was the first systematic attempt to collect and analyse global data on homicide and to investigate the link between homicide and development specifically.

The truth of violence

While each of these institutions produce knowledge about violence and security that is different in orientation and disciplinary approach, much of the knowledge that is produced by these organisations is implicitly informed by a correspondence theory of truth in which facts are understood as direct reflection of "reality" i.e. truth is a correspondence to a fact (David 2009). "Facts" reveal the truth about an objective reality.

In this paradigm we can make a claim about the universe and then prove this claim with observations and measurements. This is not to dismiss the knowledge claims made in many of the institutions that are addressing violence, but to situate these claims within a particular epistemological orientation that believes that with the correct information we can reveal the "reality of violence" (<www.unviolencestudy.org>) and formulate "scientifically-based" interventions on the basis of this reality.

Paradoxically the different international organisations dealing with violence are deeply engaged with discursive constructions around the "reality" of violence. While each of these organisations purport to capture this reality through the measurement of the extent and types of global violence, each interpret this information through different analytical lens that lead them to propose a variety of global responses and "solutions" to violence. These analyses may not be contradictory, and in some instances they are complementary, but their claims to universality of knowledge, does not reflect the enormous "disjuncture" (Appadurai, 1990) of the globalised world or even the multiplicity of discourses among those proposing global solutions to violence.

Thus the World Health Organisation argues that violence is a preventable global public health problem. The preface to the *World Report on Violence and Health* of 2002 states that it is "the first comprehensive summary of the problem [of violence] on a global scale" (Krug *et al.*, 2002). While most of the report's focus is on interpersonal violence it also argues the same public health approach can be taken to "collective violence" (Krug *et al.*, 2002). In a later report the organisation states emphatically that it has the "evidence" for "what works" (Liverpool JMU and WHO, 2010). However on a closer reading the "evidence" presented in the 2010 report, *Violence Prevention the Evidence*, does not in fact unambiguously support the strategies presented. What emerges, inevitably, are indications of both success and failure in various violence prevention interventions, despite the certitude of the framing discourse.

The Geneva Declaration Secretariat in the *Global Burden of Armed Violence Report* of 2008, produced after the signing of the *Geneva Declaration on Armed Violence* and Development, was one of the formative reports ostensibly written to provide evidence of the impact of violence on human development. The scope of the report is extensive, seeking to bring together data on direct *and* indirect deaths as a result of conflict *and* non-conflict armed violence. Like many global reports on violence it makes large claims on its own behalf, stating on its cover that it is "the first comprehensive assessment of the scope of human tragedy resulting from violence around the world" (Geneva Declaration Secretariat 2008). While the report asserts that "armed violence stunts human, social, and economic development" (Geneva Declaration Secretariat 2008: iii), it largely focuses on quantifying deaths and injury rather than exploring this link analytically. Instead it calls on states to invest in effectively monitoring "armed violence trends" and to "quantify the effectiveness" of armed violence prevention programmes (Geneva Declaration Secretariat, 2008: 6). The report concludes that on the basis of "up-to-date data and research" "armed violence is preventable". A follow-up study conducted by the Geneva Declaration Secretariat in 2011, *Global Burden of Armed Violence 2011*, provides a much more systematic and detailed analysis of the relationship between armed violence and development progress as measured by the Human Development Index (HDI) and the Millennium Development Goal (MDG) indicators. Unsurprisingly the report finds that "lethal violence constrains development progress" (Geneva Declaration Secretariat, 2008: 9). While noting an inverse relationship between homicide and HDI, i.e. those countries with a higher HDI are likely to experience less homicide, the report also concludes that it is difficult to establish that violence causes lower HDI scores. It nevertheless concludes that its findings "confirm the body of research that identifies a robust relationship between income inequality and violent criminality" (Geneva Declaration Secretariat, 2008: 10).

The World Bank in the *World Development Report* states emphatically that violence is the "primary development challenge of our time" and argues that "no low-income fragile or conflict-affected country has yet to achieve a single United Nations Millennium Development Goal" (2011: 1). According to the report while the forms of violence have shifted away from major inter-state wars, new forms of conflict and violence pose equal challenges to development. Many countries are caught in "vicious" and "repeated cycles of violence" (2011: 6) that impact on GDP and economic growth over an extended period. For the World Bank legitimate institutions are the key to helping societies cope with the stresses of unemployment, economic inequality and rapid urbanization without resorting to violence. The report identifies a "roadmap for breaking cycles of violence at the country level" (2011: 8). While arguing that is does not want to provide any "cookbook" (2011: 16) solutions, the report identifies "four core tools" (2011: 16), based on its interpretation of the evidence and lessons learnt, that it argues can be applied in the context of "fragile" (2011: v) states to break repeated cycles of violence. These include "restoring confidence" in the political system, "transforming institutions", "national action to address external stress" and "feasible results indicators to demonstrate overall progress" (2011: 8).

Like *the Global Burden of Armed Violence Report* UNODOC's 2011 *Global Study on Homicide* also seeks to bring together extensive sets of data including, "global, regional, national and subnational homicide data"

(2011: 5). It makes an explicit link between homicide and development, implicitly building on the conceptual framework first established by the *Geneva Declaration on Armed Violence* and elaborated in subsequent global reports on violence. The report argues that homicides are highest in those countries with low human and economic development and high levels of inequality (2011: 10). It therefore states that, "the development agenda should include crime prevention policies" (2011: 5). The report also links development to the "rule of law" (2011: 5), arguing that the rule of law is a "vital piece in the human development puzzle", which is critical to establishing effective democratic governance and is associated with reduced homicide rates.

Thus each of these reports claim to present a globally applicable "diagnosis" of the problem of violence and a set of "tools" to address the issue based on claims to know "what works" in local contexts around the world. In an era of globalisation it appears appropriate that violence and problems of security should be addressed as global issues when global networks between countries are conduits for multiple exchanges of "goods" (food, consumer goods, drugs and people). The question is, is this an equal global partnership to address violence?

Instead interconnectedness can lead powerful global actors to directly intervene in local conflicts, fearing the export of "terror" along the networks that link country to country or believing that solutions for violence formulated by global institutions can be uncritically applied in local contexts. The consequences of international intervention have as a result often been deeply ambiguous for citizens, either unleashing new forms of conflict in places such as Iraq or failing to effectively address violence as a result of the transfer of international models of violence prevention that may or may not be appropriate in local contexts. For example the application of the "community policing" model to South Africa (see Brogden, 2004). The limitations of these interventions are, however, presented as the consequence of the limitations of the countries in which they have been implemented, not a result of the nature of the intervention itself. As the World Bank asks, "How is it that, almost a decade after renewed international engagement with Afghanistan, the prospects of peace seem distant?" (2011: 2).

It is unclear therefore whether this recognition of interconnection includes not only a recognition of how "undeveloped" countries can contribute to lack of security in more "developed" countries but also how this relationship works in reverse i.e. the multi-directional nature of the transfer of violence.

As Comaroff and Comaroff have argued the crime and disorder of the developing world is profoundly linked to global historical processes, most notably colonialism and more recently globalisation. The spread of neo-liberal economics has profoundly deepened inequality both within states and between states in the global order (Comaroff and Comaroff, 2006). Trade in the global centre gains significantly by exporting its production to the global south or extracting minerals at minimal cost with little investment in the social economies of the developing world. These forms of "violence" do not appear in the reports that purport global coverage of violence. At the same time the "prescription" of democracy and liberalism has not had unambiguously positive consequences in post-colonial contexts as the liberalisation of economies and societies has in some instances led to forms of deregulation, which have created the space for a "proliferation" of lawlessness, as well as the development of international criminal economies that link the developed and developing world in a web of illegality that apes the logic of supply and demand in the formal global economy (Comaroff and Comaroff, 2006).

The *Global Report on Homicide* acknowledges exactly these links when it states that, "A major cause of violent crime in Central America is also the region's strategic location between the lucrative cocaine consumer market in Northern America (although the European market is increasingly important) and the main areas of coca cultivation in Colombia, Peru and the Plurinational State of Bolivia" (UNODC, 2011: 51).

On the other hand, the World Development report, implicitly addressing its "northern" audience, writes, "For those who now live in more stable neighborhoods, it may seem *incomprehensible* how prosperity in high-income countries and a sophisticated global economy can coexist with extreme violence and misery in *other* parts of the globe" (The World Bank, 2011: 1, own emphasis). The report then contrasts the pre-modern African with the modern nation-state of the West:

> The pirates operating off the coast of Somalia who prey on the shipping through the Gulf of Aden illustrate the paradox of the existing global system. How is it that the combined prosperity and capability of the world's modern nation-states cannot prevent a problem from antiquity? (The World Bank, 2011: 1).

Thus these "pirates" appear with no history, context or agency. What in the conditions in Somalia that may have shaped their conduct is opaque, how "piracy" may be linked to the very modern processes of globalisation and transnational flows of power and goods across the globe, is left unproblematised. Instead these "pirates" appear merely as evidence of the continuation of "antiquated" forms of living, agency and subjectivity in Africa.

There is in fact little that is "incomprehensible" about the co-existence of a "sophisticated global economy" and the "violence and misery" that characterize life in a large proportion of the world. These phenomena are in fact inter-related parts of the same global form and are fundamental to the current global order of power in which various forms of structural violence make "a certain kind of domestic peace possible" (Scheeper-Hughes and Bourgois, 2004: 20).

Theorists working within the development paradigm in the global south have critiqued some of these forms of knowledge production. Writers such as Robinson (2002) and Parnell (1997) have therefore argued for the importance of seeing the globe in terms of the existence of diverse, parallel and interlinked modernities characterized by multiple rationalities, forms of ethics and conceptions of citizenship, rather than organising the world into a hierarchy of the developed and the non-developed. Roy has asserted the need to "dislocate the Euro American centre of theoretical production" (Roy, 2009: 820) in order to, as Comaroff and Comaroff have recently argued, develop "theory from the south" (Comaroff and Comaroff, 2012; see also Parnell, 1997; Mbembe and Nuttall, 2004; Robinson, 2006; Mbembe, 2012; Pieterse, 2012).

The challenge therefore is not to simply collect increasingly extensive datasets on violence and injury, but to fundamentally reconfigure our understanding of the world in ways that acknowledge the forms of modernity that exist in the global south, in order to achieve more socially just and less violent forms of life throughout the globe.

Securing the "developed" world?

Global concerns and anxieties about security have to be located in the broader paradigm of a preoccupation with, and organisation around, "risk" in late modernity, particularly in the "developed" world. A risk *society*, as articulated by Beck, is a society organised around the calculation and prevention of risk or harm (1992). This is not only the calculation of "objective" risk but is as fundamentally about subjective perceptions of risk, the communication of risk and the social experience of living with risk (Ekberg, 2007).

Thus as Giddens points out, the "Preoccupation with risk in modern social life has nothing directly to do with the actual prevalence of life-threatening dangers" (Giddens, 1991: 115). In fact in the developed world people are in a much more secure position than in the past, relatively safe from the contingencies of uncontrolled epidemics, starvation, and with access to the basic necessities of life. What has changed in these societies in late modernity is that they are fundamentally organised around the notion of risk, exemplified by the rise of the insurance industry, which is premised on the calculation of risk and protection against future misfortune. In this context the subjective construction of risk becomes a "virtual" (Beck, 2000: 213) and unstable reality, "where risks are believed to be real the foundations of business, politics, science and everyday life are in flux" (Beck, 2000: 214). Thus in reflexive modernity the fear of global insecurity increases and risk assessment becomes "a more or less ever-present exercise, of a partly imponderable character" (Giddens, 1991: 124).

A critical prism through which the "risk" of violent conflict has been framed in the last ten years, is the thesis that security and development are integrally linked. It was the 2006 *Geneva Declaration on Armed Violence and Development*, that brought the conceptual link between security and development to the forefront of the policy debate in recent years (see also OECD, 2004; US Department of State and USAID, 2003; EU, 2008). By the time the World Bank published its 2011 *World Development Report* subtitled, *Conflict, Security and Development*, the link between security and development had become axiomatic in the policy domain despite contestation in academic literature about the empirical validity of this claim.

The idea that security is linked to development is not a new concept and in fact dates back to the colonial period. However its most recent articulations relate to post 9/11 anxieties about the threat of terrorism from "undeveloped" countries. In the US development and security have been explicitly linked since the post-World War Two period with, for example the Marshall plan, which was a US initiative to provide development aid to countries in Europe in order to ensure that they did not become vulnerable to Communism (Bueger and Vennesson, 2009).

This association between development aid and security continued to play a significant role in the international arena, particularly during the early Cold War period, when development aid to the "third world" was seen as closely linked to the security of developed countries in the Western, anti-Communist bloc of nations. In 1951 the US set up the US Mutual Security Agency (MSA) which was overtly designed to integrate military and economic programmes (Beall *et al.*, 2006). From the 1960s other countries joined the US as major donors to the "third world", coordinated through Organisation for Economic Co-operation and Development (OECD) established in 1961 (Bueger and Vennesson, 2009). However, from the 1970s onwards there was a growing understanding of the importance of focusing on development for development's sake rather than as an instrumental means to achieve "security" (Beall *et al.*, 2006). Multilateral institutions, rather than individual states, played an increasingly important role in relation to development (Bueger and Vennesson, 2009) so that the development could be less directly tied to specific state interests, although donor countries continued to run their own development initiatives.

Post 9/11 and the US's declaration of a "War on Terror" there appears to have been a reversal in efforts to distinguish development from narrow security interests, including, according to Beall, in both bilateral and multilateral institutions (Beall *et al.*, 2006). This has led to what Stern & Öjendal (2010) have called the "securitisation" of development. In this approach underdevelopment is seen as "dangerous" (Geneva Declaration Secretariat, 2008: 63), not to the inhabitants of societies where development is not taking place, but to the "international" community. To respond to this "threat" the Geneva Secretariat argues that donor country investment and interventions are necessary "to bolster governance and growth opportunities in so-called fragile or weak states" (Geneva Declaration Secretariat, 2008: 63).

Thus, the risk paradigm when applied to the domain of the social, inscribes fear into the heart of the ways in which the world is known. Instead of grappling with new imaginaries of freedom the risk paradigm focuses attention on immediate security and safety in a world now configured as a place of danger and imminent threat. From this vantage point creative and inventive forms of engagement, theorizing, doing and dreaming are stifled in a world of fear and closure.

Hudson has therefore argued that that the growing interdependency and perceived "risk" of the global political system has challenged the liberal goal of balancing security and freedom through the social contract, leading to novel forms of governmentality and social control (2003: 40).

Thus a number of authors have noted that the current focus on the relationship between security and development appears to subordinate development needs to the security interests of countries in the West (Duffield, 2001; Beall, 2006; Buur, Steffen and Stepputat, 2007; Bueger and Vennesson, 2009). Chandler however, cautions about the need to distinguish between policy rhetoric and actual coherence at policy or programme level (2007). Nevertheless the international articulation of the link between violence and development, paradigmatically represented by UN Secretary-General Koffi Anan's assertion in 2005 that, "In an increasingly interconnected world, progress in the areas of development, security and human rights must go hand in hand. There will be no development without security and no security without development," has had a significant impact on the actual flow of development aid to the developing world, with the major beneficiaries of Official Development Assistance (ODA) going to countries that could be seen to be the most potential security threat to the US, namely Iraq, Afghanistan, Sudan, Egypt and Ethiopia (Giovannetti, 2010). In the European Union resources allocated to development aid have almost doubled since the beginning of the 2000s (Carbone, 2009) to respond to these anxieties about the "threat" of underdevelopment.

Despite repeated assertions in the policy arena that the link between security and development is self-evident, academic work that has sought to investigate this link both conceptually and empirically has noted the limitations of this discourse. Some have focused on the potentially detrimental consequences for development of reducing the entire universe of development aid to countries which are seen as a potential "terrorist" threat (Beall *et al.*, 2006) as well as the shaky empirical grounds of linking terrorism to poverty, "failed" or "weak" states (Hehir and Patrick, 2007). A number of authors have noted the lack of conceptual clarity (Venesson, 2009; Stern and Öjendal, 2010) or the contested nature of the concepts of development and security (Spear and Williams, 2012), which make it difficult to draw a clear conceptual or empirical link between the two (See Tschirgi *et al.*, 2010; Amer *et al.*, 2012; Spear and Williams, 2012). They note the bluntness of the current articulation of the link, which does not differentiate between different phases of conflict or levels of policy (Tschirgi *et al.*, 2010) and point to the difficulties of understanding whether conflict is an outcome or cause of underdevelopment. Spear and Williams conclude that, "it is impossible to draw meaningful conceptual generalisations across different times and places" (2012: 21).

Colonising the future

The inscription of a relationship between security and development within risk discourse leads to a regime of risk classification oriented towards maintaining cultural coherence through stereotypical articulations of "us" and "them" (Ericson and Haggarty, 1997: 287) facilitated by false constructions of group homogeneity. Such a system is guided by instrumental rationality aimed at increasing the governability of diverse populations in ways that reduce complex social problems to a-historical technical phenomena that can be resolved through "rational", utilitarian action divorced from moral subjectivity or affect.

The global institutions in which most reports around violence have been generated, are deeply inscribed in this logic. The projection of risk in these global reports creates an illusionary perspective, in which the "outward" identification and categorisation of "risky" societies from a position of "stability, non-violence and modernity" becomes normalised. The "developing world", "global south", "third world" become the object of a gaze from afar, the focus of a risk calculus based on surveillance and risk classification. Countries across the globe are "ranked" and categorised according to their dysfunctions. Maps generated from these exercises show swathes of blood red across whole continents and regions, reports are littered with photographs of the dead and suffering. (See Geneva Declaration Secretariat, 2008, World Bank Datasets on Violence,

WHO Global Health Observatory (GHO)). This is not to say that these countries are not violent or poor but to ask what lies beneath those bloody maps, what are the local and global, contemporary and historical socio-political processes, which have led to these conditions and where are the voices of those located in these zones of disorder? How or do they shape this global knowledge production about the conditions of their lives?

While many global reports pay rhetorical attention to specificity and local context, in fact this "context" generally appears as sites where global models and approaches need to be adapted and modified, rather than sites which may profoundly contest the basis of the knowledge production on which these global models are founded. Conditions in the global south become the basis, not for an interrogation of the complexity of the problem of violence in such contexts but evidence of unequivocal "failure" (See The Fund for Peace *Failed State Index* <http://ffp.statesindex.org>) against a normative model of the state and Western modes of living, posed in opposition to archaic forms of "their own culture" (Klug *et al.*, 2002, p: xxii). However the deeply cultural nature of the "roadmaps" (The World Bank, 2011: v) that are prescribed for these societies are presented as scientific neutrality.

For example at the heart of the WHO's public health model is the sovereign, context transcending subject, able to choose their ends and means despite historical and social contexts. This subject is ultimately a fiction based on a long history of liberal Western contractarian thought that ignores the deeply limited possibilities for individual choice in contexts profoundly shaped by significant histories of structural and explicit violence. It also does not engage with the many communitarian forms of solidarity, agency and social regulation, which remain salient in many post-colonial contexts.

The prescriptions for the "rule of law" and "legitimate" institutions by organisations such as UNODOC and the World Bank, which are based on the liberal democratic model of the state are founded on a valid concern with curtailing the arbitrary exercise of power. However, recommendations for the "rule of law", often fail to engage meaningfully with the deeply ambiguous history of the state and of law itself in many post-colonial contexts where the norm as well as the rule of law remain deeply contested. As Krygier points out a critical condition for the realisation of the "rule of law" is that "institutionalised norms need to *count* as a source of restraint and a normative resource, usable and with some routine confidence used in social life" (2007: 13, original emphasis). Therefore establishing the norm of law in a context where the law has been deeply entangled with the colonial project and, which even in the post-colonial context does not generally count as a normative resource, is an extraordinarily complex task that does not lend itself to generic models.

A critical presumption of the risk discourse that shapes many global reports on violence, is the possibility of instrumental control over the future, what Giddens calls an increase in attempts to "colonise" the future (Giddens, 1991; Beck, 1992), to "bring it under our dominion" (Giddens, 2000: 26).

Therefore in this paradigm, with enough information, the future can be made transparent. We can "predict" the outcome if we list enough risk factors and collect enough data. One of the key features of risk society is the reflexivity about risk and harms that have emerged from modernity (i.e. environmental pollution, terrorism, increased crime and disorder in urban centres). This leads to increasing demand for and dependency on expert systems and technologies to increase knowledge of future risks, which results in the paradox of risk society, i.e. that concentrating on risk and security solutions drives the need for more knowledge of risk but "this search for inexhaustibly detailed and continuous risk management knowledge gives rise to new knowledge of insecurity requiring more and more technical solutions to constantly proliferating risks" (Ericson and Haggerty, 1997: 85).

This anxiety to accumulate technical knowledge about the risk of violence is evident in the significant number of global reports published over the past decade. The 2006 *Geneva Declaration on Armed Violence and Development* committed signatories to "support initiatives to *measure* the human, social, and economic costs of armed violence", "to assess *risks* and vulnerabilities" and "to achieve *measurable* reductions in the global burden of armed violence" by 2015 (Geneva Declaration Secretariat, 2008: vii own emphasis).

In order to do this, the 2008 *Global Report on the Burden of Armed Violence* asserts that national and international organisations must "draw on social science and public health methods to quantify the effectiveness of armed violence prevention and reduction programmes" (Geneva Declaration Secretariat, 2008: 6). Armed violence is therefore presented as primarily a data challenge. Context, history and power fade from view. This violence may have "political implications" (Geneva Declaration Secretariat, 2008: 1), which lead some actors to understate or conceal the problem. This can make data collection "difficult" and "controversial", nevertheless "impartial analysis" based on "reliable" data will enable "practical" measures to be implemented to reduce armed violence (Geneva Declaration Secretariat, 2008: 1).

Public health approaches and risk analysis produce ostensibly neutral but deeply atavistic "indicators of states at risk of collapse and internal conflict" which predict breakdown, when, for example, the "ethnic composition of the ruling group [is] sharply different from that of the population at large" (Klug *et al.*, 2002: 220). The vastly different role that ethnicity has played in various sites of conflict from Yugoslavia to Sierra Leone, its relative significance in relation to the many other factors that lead to conflict and its historical, political and economic conditions of emergence in particular contexts, are all reduced to an indicator transparently extrapolated from the most reductionist understanding of the conflict in Rwanda, which is then presented as universally predictive of conflict in countries across the globe.

The "lacuna" of data in large parts of the world, noted in many of these global reports, is presented as merely part of the data challenge that prevents the development of effective technical interventions. The *Global Report on Homicides* devotes an entire chapter to "Data challenges" (UNODOC, 2011: 83). The *World Development Report* (2011) notes the "the lack of available quantitative and qualitative data, due to challenges of security and access, along with low statistical capacity" (The World Bank, 2011: xix). As a result "countries most affected by violence often register empty data columns" (The World Bank, 2011: xix). This has led to anomalies in global rankings, which, for example, place the Bahamas at a higher rate of violence than Afghanistan (UNODOC, 2011).

Therefore in this paradigm, we can remake the future, ideally, in "our" own image. According to the WHO we do not have to accept violence as an "inevitable part of the human condition" (Klug *et al.*, 2002: 3) instead, "violence can be prevented and its impact reduced, in the same way that public health efforts have prevented and reduced pregnancy-related complications" (Krug *et al.*, 2002: 3). Therefore, "The factors that contribute to violent responses – whether they are factors of attitude and behaviour or related to larger social, economic, political and cultural conditions – can be changed… This is not an article of faith, but a statement based on evidence" (Krug *et al.*, 2002: 3). Rational choice will therefore vitiate the complexity of social and structural causation and the contingency of individual decision.

Thus the vision of the future to be remade is deeply sanitised, shorn, for the WHO, of the "culture" and "tradition" of the "other" who is posed in opposition to *Homo Economicus*, the rational-economic actor of the West. The WHO argues, "Cultural norms must be dealt with sensitively and respectfully in all prevention efforts – sensitively because of people's often passionate attachment to *their* traditions, and respectfully because culture is often a source of protection against violence" (Krug *et al.*, 2002: 16 own emphasis). Therefore, "Experience has shown" that these cultural others and their traditional leadership must be consulted (Krug *et al.*, 2002: 16) in order to ensure public health approaches that are culture-free and "based on the rigorous requirements of the scientific method" (Krug *et al.*, 2002: 4).

All this is not to argue that trying to collect information on violence is not critically important, particularly in the global south where formal systems of documentation are often fragile or non-existent. However, this documentation must make it possible to increase rather than reduce complexity in order to understand the many dynamics of violence, its historical conditions of emergence, its current political economy, the impact of policy, of political struggle, of individual motivation and of social relation. However, the form of knowledge that is being pursued, with its almost complete emphasis on quantification is problematic because it cannot critically engage with the prior conditions of its own knowledge production, i.e. the sets of assumptions on the basis of which choices regarding which "data" is collected, are implicitly made. The focus on quantification means that stereotypes and assumptions about the contexts where violence is happening remain unexamined, unchallenged by actual forms of living in different contexts.

Therefore we need to understand the rationalities which underlie this appetite for data, the nature of the hopes that are embedded in these efforts, the specificity of the future they envision and the limitations within which they are inscribed.

Conclusion

The wave of global reporting and data collection around violence that has characterised the past decade, makes an extraordinary breadth of claims to knowledge, to universality, to the ability to represent the "reality" of violence in a dizzying display of quantified death drawn from global data on "inter-personal" violence, "self-directed" violence, "collective" violence, "armed violence", "conflict-related" violence, "non-conflict" related violence, on "homicide", on "gender violence", on "security and violence", on "public health and violence" and the development of an equally dizzying array of "tool-kits", "road-maps" and other recommendations. The problem is not, as some have argued, merely a lack of "policy coherence" (Carbone, 2008) but a fundamental conceit around total knowledge and underlying this a far more pragmatic concern with safety "at home", which is revealed by the actual flows of development aid. This is not to undermine the extraordinary work by many people working in institutions around the world to alleviate violence and suffering. However, it does ask us to pause and ask whether "policy fixes" such as institutional reform can address conditions that may exist "beyond civility" and beyond the prescriptions of Weberian rationality (Pieterse, 2010).

This paper has only begun to point to some of the ways in which we may interrogate the knowledge claims of global institutions working with violence. Implicit in much of this discourse are deeply engrained assumptions about the nature of social and political life founded on the historical and contemporary experience of countries in the global north where many of these reports are conceived and generated. As a result interventions around violence are oriented to recreating societies which mirror these familiar forms, despite the specificity of the historical conditions of emergence of the societies where globalised interventions are implemented. This is problematic not only because of the hierarchical relationships that are created between ostensible zones of peace and modernity and those of violence and pre-historicity, but as importantly because prescribing solutions which dismiss as cultural alterity existing forms of social and political life, are likely to have little impact in actually reducing violence.

Thus while we may have a normative development agenda, this needs to be pursued, as Sen has argued, in order to maximise and expand "the freedoms that people enjoy", to create a world that enables human beings to develop those "capabilities" necessary for them to realise "the good life" as they value and envisage it (Sen, 1999). The development agenda cannot be wielded as a normative stick against which to judge the lack and the threat in the "other". The critical challenge is *how* we bring an agenda for the "good life" into being in full cognisance and recognition of the multiplicity of ways in which to realise this good life and the material and social conditions which prevent this realisation. It is only on a terrain of epistemological equality that meaningful and ethical interactions can take place in order, indeed, to meet our aspirations for social justice and substantive "freedom". In this engagement we need to fashion new conceptual frames that are adequate to the task of analysing historical and emerging forms of violence in ways that do not attempt to reduce its complexity and variability to simplistic causalities or aver definitive and universal "solutions".

References Cited

Amer, R., Swain, A., Öjendal, J. (2012). *The Security-Development Nexus: Peace, Conflict and Development*. London: Anthem Press.

Appadurai, A. (1990). "Disjuncture and Difference in the Global Cultural Economy" *Theory Culture Society*, 7, pp. 295-310.

Annan, K. (2005). *In Larger Freedom: Towards Development, Security and Human Rights for All*. Report of the Secretary-General. New York (UN A/59/2005), annex, point 2.

Beall, J., Goodfellow, T., Putzel, J. (2006). "Introductory Article: On the Discourse of Terrorism, Security and Development" *Journal of International Development* 18, pp. 51-67.

Beck, U. (1992). *Risk Society: Towards a New Modernity*. London: Sage.

Beck, U. (1999). *World Risk Society*. Cambridge: Polity.

Beck, U. (2000). "Risk Society Revisited: Theory, Politics and Research Programmes". In Adam, B. Beck, U., Van Loon, J. (eds.). *The Risk Society and Beyond: Critical Issues for Social Theory*.

Bellis, M., Jones, L., and Hughes, K. (2010). "Preventing and Reducing Armed Violence: What Works?" *Background Paper for the Oslo Conference on Armed Violence*. New York: UNDP.

Brogden, M. (2004). "Commentary: community policing: A panacea from the West" *African Affairs*, 103/413, pp. 635-649.

Bueger, C. and Vennesson, P. (2009). *Security, Development and the EU's Development Policy*. European University Institute, Robert Schuman Centre for Advanced Studies, April. Available at <http://erd.eui.eu/media/vennesson2.pdf>.

Buur, L., Jensen, S., Stepputat, F. (eds.) (2007). *The Security-Development Nexus: Expressions of Sovereignty and Securitization in Southern Africa.* Nordiska Uppsala: Afrikainstitutet & Cape Town: HSRC Press.

Carbone, M. (2008). "Mission Impossible: The European Union and Policy Coherence for Development" *Journal of European Integration*, 30 (3), pp. 323-342.

Carbone, M. (2009). *Desperately Seeking Policy Coherence: Aid and Security in the Development Policy of the European Union.* Paper Prepared for the Workshop "Transforming Political Structures: Security, Institutions and Regional Integration Mechanisms". European Report on Development in Florence, Italy, 16-17 April.

Chandler, D. (2007). "The security-development nexus and the rise of 'anti-foreign policy'" *Journal of International Relations and Development*, 10, pp. 362-386.

Comaroff, J. & Comaroff, J.L. (2004). "Criminal Obsessions, after Foucault: Postcoloniality, Policing, and the Metaphysics of Disorder" *Critical Inquiry*, 30 (4), pp. 800-824.

Comaroff, J. & Comaroff, J.L. (eds.) (2006). *Law and Disorder in the Postcolony*. Chicago and London: University of Chicago Press.

Comaroff, J. & Comaroff. J.L. (2011). *Theory from the South: or how Euro-America is evolving towards Africa.* Colorado: Paradigm Publishers.

Das V. Kleinman, A. Ramphele, M. & Reynolds, P. (eds.) (2000). *Violence and Subjectivity*. Los Angeles: Berkley & London: University of California Press.

David, M. (2009). *The correspondence theory of truth, Stanford Encyclopedia of Philosophy* (Winter 2009 Edition). Edward N. Zalta (ed.), URL <http://plato.stanford.edu/entries/truth-correspondence/>.

Duffield, M. (2001). *Global Governance and the New Wars. The Merging of Development and Security*. London: Zed Books.

Ekberg, M. (2007). "The Parameters of the Risk Society: A Review and Exploration" *Current Sociology*, 55 (3), pp. 343-366.

Ericson, R.V. & Haggerty, K.D. (1997). *Policing the Risk Society*. Oxford: Oxford University Press.

European Council (2008). *Report on the Implementation of the European Security Strategy – Providing Security in a Changing World*. Brussels: European Council.

Foucault, M. (1984). "Nietzsche, Genealogy, History", 1971. In Rabinow P. (ed.). *The Foucault Reader.* New York: Pantheon Books.

Geneva Declaration Secretariat (2008). *Global Burden of Armed Violence*. Geneva.

Geneva Declaration Secretariat (2008). *Global Burden of Armed Violence 2011*. Geneva.

Giddens, A. (1991). *Modernity and Self-Identity: Self and Society in the Late Modern Age*. Cambridge: Polity Press.

Giddens, A. (2000). *Runaway World: How Globalization is Reshaping Our Lives*. New York: Routledge.

Giovannetti, G. *The Security-Development Nexus: How to Bridge the Gap between Foreign Security Policies and Development Policies, Introductory notes*. European University Institute, Robert Schuman Centre and University of Firenze. Presentation at New York, Peacebuilding and the Security-Development Nexus April 26th. Available at <http://www.iss.europa.eu/uploads/media/Peacebuilding>.

Hehir, A. (2007). "The Myth of the Failed State and the War on Terror: A Challenge to the Conventional Wisdom" *Journal of Intervention and Statebuilding*, 1 (3), pp. 307-332.

Hudson, J. (2003). *Justice in the Risk Society: Challenging and Re-Affirming 'Justice' in Late Modernity*. London: SAGE Publications.

Krug E.G. *et al.* (eds.) (2002). *World Report on Violence and Health*. Geneva: World Health Organization (WHO).

Krygier, M. (2007). *The Rule of Law: Legality, Teleology, Sociology*. University of New South Wales Faculty of Law Research Series Paper 65.

Liverpool John Moores University (JMU) & World Health Organisation (WHO) (2010). *Violence Prevention: the Evidence*. Geneva: WHO Press.

Mbembe, A., and S. Nuttall (2004). "Writing the World from an African Metropolis" *Public Culture*, 16 (3), pp. 347-72.

Mbembe, A. (2012). "At the centre of the knot" *Social Dynamics: A journal of African studies*, 38 (1), pp. 8-14.

OECD/DAC (2004). *The Security and Development Nexus: Challenges for Aid*. Paris: OECD, DAC.

Parnell, S. (1997). "South African Cities: Perspectives from the Ivory Tower of Urban Studies" *Urban Studies*, 34 (5-6), pp. 891-906.

Patrick, S. (2007). "'Failed' States and Global Security: Empirical Questions and Policy Dilemmas" *International Studies Review* 9, pp. 644-662.

Pieterse, E. (2010). *Cityness and African Urban Development.* Working Paper No. 2010/42, United Nations University, UNU-WIDER, World Institute for Development Economics Research.

Pinheiro, P.S. (2006). *World Report on Violence against Children.* Secretary-General's Study on Violence against Children. Geneva: United Nations.

Rabinow, P. (ed.) (1984). *The Foucault reader.* New York: Pantheon Books.

Robinson, J. (2002). "Global and World Cities: A View from off the Map" *International Journal of Urban and Regional Research*, 26 (3), pp. 531-554.

Roy, A. (2009). "The 21st-Century Metropolis: New Geographies of Theory" *Regional Studies*, 43(6), pp. 819-830.

Sen, A. (1999). *Development as Freedom.* Oxford: Oxford University Press.

Scheeper-Hughes, N. & Bourgois, P. (eds.) (2004). *Violence in War and Peace: An Anthology*. Maldon, Oxford and Carlton: Blackwell Publishing.

Spear, J. and Williams, P.D. (2012). *Security and Development in Global Politics: A Critical Comparison*. Washington, DC: Georgetown University Press.

Stern, M. & Öjendal, J. (2010). "Mapping the Security-Development Nexus: Conflict, Complexity, Cacophony, Convergence?" *Security Dialogue*, 41(1), pp. 5-30.

The World Bank (2011). *World Development Report 2011: Conflict, Security and Development*. Washington DC.

Tschirgi, N., Lund, M.S., Mancini, F. (2010). *Security and Development: Searching for Critical Connections*. Boulder: Lynne Rienner Publishers.

UNICEF (United Nations International Children's Emergency Fund) (2014). *Hidden in Plain Sight: A Statistical Analysis of Violence Against Children.* UNICEF.

United Nations Office on Drugs and Crime (UNODC) (2011). *Global Study on Homicide: Trends, Context, Data*. Vienna: United Nations Office on Drugs and Crime (UNODC).

US Department of State and USAID (2003). *Security, Democracy, Prosperity: Strategic Plan, Fiscal Years 2004-2009, Aligning Democracy and Development Assistance*. Washington DC: USAID.

Willman, A. & Makisaka, M. (2011). *Interpersonal violence prevention: A Review of the Evidence and Emerging Lessons*. World Development Report 2011 Background Paper.

World Health Organisation (WHO) (2013). Department of Reproductive Health and Research, London School of Hygiene and Tropical Medicine & South African Medical Research Council. *Global and Regional Estimates of violence against women: prevalence and health effects of intimate partner violence and non-partner sexual violence*. Geneva: WHO Press.

Aportes de la televisión a la formación ciudadana en el contexto de las violencias del sistema de salud en Colombia[1]

Claudia PUERTA SILVA & Alejandro AGUDELO CALLE

Resumen

El objetivo de la presente reflexión es identificar cómo la televisión colombiana contribuye con información en salud y, desde esta perspectiva, qué tanto y cómo está aportando a la formación ciudadana de los televidentes. Reconociendo que la televisión podría tener un potencial formativo, este trabajo exploró, además, los aspectos que condicionan la recepción de contenidos audiovisuales y su eventual apropiación en ejercicios ciudadanos. La investigación fue realizada entre marzo y octubre de 2012 y los resultados presentados son exploratorios, descriptivos y cualitativos. Para analizar los contenidos, se realizó un análisis textual de cuatro programas emitidos en un lapso de 15 días, para un total de 21 emisiones. La recepción televisiva se analizó a partir de 30 entrevistas semiestructuradas a televidentes, productores y presentadores y un grupo focal. Se identificó que los contenidos televisivos se dirigen hacia la enfermedad y su prevención y no abordan suficientemente ni el funcionamiento ni la estructura del sistema de salud y menos los derechos y deberes. Tampoco aportan a la movilización política o a la participación ciudadana con respecto a los derechos a la salud. Con relación a la recepción de contenidos televisivos por parte de los

[1] Ponencia presentada al panel 5: La percepción del Estado desde la ciudadanía: teorizaciones de las comunidades frente a la capacidad del Estado para garantizar su salud y bienestar en contextos de violencia y conflicto. Conferencia RISC 2012, Ciudad de México D.F., 30 de octubre al 2 de noviembre 2012. Resultados del proyecto "Televisión y su aporte a la configuración de ciudadanías en salud", financiado por el Comité para el Desarrollo de la Investigación CODI, el Instituto de Estudios Regionales INER y la Facultad de Ciencias Sociales y Humanas de la Universidad de Antioquia y avances de la tesis doctoral de Alejandro Agudelo Calle. Se agradece a la Estrategia de Sostenibilidad 2013-2014 de la misma Universidad. Agradecemos al investigador Aléxander Hernández Marín y a los estudiantes que participaron en el proyecto de investigación por sus aportes: Yesenia Jaramillo, Marilly Rendón y Alexis Hernández. También agradecemos las sugerencias realizadas por quienes evaluaron el manuscrito.

televidentes, se identificó que la vulneración constante del derecho de los ciudadanos al acceso a los servicios de salud condiciona la apropiación de los mensajes y la acción ciudadana. Las barreras a tales servicios se denominan en este artículo violencias del sistema de salud. Esta forma de violencia, considerada aquí como una mediación, se potencia debido a la desinformación de los ciudadanos sobre sus derechos y deberes y sobre el funcionamiento del sistema de salud.

Palabras clave

Televisión, ciudadanías en salud, derechos, recepción televisiva, violencias del sistema de salud, Medellín, Colombia.

*

Introducción

La televisión ha sido señalada por su potencia educativa como una herramienta vital en la formación en salud de los ciudadanos (García Reina, 2004; Kubey & Csikszentmihalyi, 1990; Singhal & Rogers, 1999). Se ha establecido que las campañas televisivas han logrado concientizar sobre un problema de salud, inducir el cambio de actitudes y conductas y promover la búsqueda de información (Rideout, 2008). La relación entre la televisión y la salud ha sido abordada por los estudios de comunicación para la salud, desde la perspectiva de la promoción de la salud (Parra & Oliva, 2004); la transmisión de mensajes de salud en la televisión pública (COMFER, 2007); y el tratamiento de la salud pública (Obregón Gálvez, Vega Casanova, & Lafaurie Molina, 2003).

En Medellín, la mayor parte de las personas tienen acceso a la televisión (Comisión Nacional de Televisión, 2010) y ve televisión (DANE, 2011), adquiriendo de ella información sobre el Estado, sus instancias, las noticias y, eventualmente, sobre sus derechos y deberes. La presencia masiva de la televisión en la vida cotidiana indica como afirma Silverstone, que "los medios se están convirtiendo en algo ambiental [...] pues están estrecha y dialécticamente entrelazados con lo cotidiano. Dependemos de los medios para llevar adelante nuestra vida diaria, y los medios se han convertido en el elemento sine qua non de lo cotidiano" (2007, p. 19).

Este artículo pretende identificar cómo la televisión colombiana contribuye con información sobre salud y, como consecuencia, al proceso de formación ciudadana. El propósito es evidenciar la potencia formativa y educativa de la televisión en temas públicos como el de la salud, especialmente en un contexto de vulneración de derechos, bajo dos supuestos: el primero, la televisión es hoy una fuente central de información y debate público en sociedades industrializadas, modernas y urbanas; y el segundo supuesto es que con contribuciones significativas de su parte, los ciudadanos tendrían más herramientas para proteger y exigir sus derechos en salud en el contexto particular colombiano. Se identificó que para los televidentes colombianos las experiencias dramáticas vividas en el sistema de salud se convierten en una mediación que condiciona la apropiación y aprovechamiento de la información aportada por la televisión para su ciudadanía.

Luego de su reforma en 1993, en el sistema de salud colombiano conviven la lógica de los derechos a la seguridad social en salud y la lógica privada de aseguramiento, oferta y demanda de servicios de salud (Molina & Ramírez, 2013). Desde entonces se ha identificado que en el sistema de salud se vulneran reiterativamente los derechos de los ciudadanos a la salud, especialmente el del acceso a los servicios de salud (Abadía Barrero & Oviedo Manrique, 2010; Agudelo Calderón, Cardona Botero, Ortega Bolaños, & Robledo Martínez, 2011; Vargas J. & Molina M., 2009; Yepes & Ramírez, 2007). Adicionalmente, se ha establecido que las barreras al acceso a los servicios se potencian y, en consecuencia, generan mayores vulneraciones cuando los ciudadanos no conocen sus derechos y deberes y no conocen el funcionamiento del sistema de salud (Agudelo Calderón *et al.*, 2011; Dover & Puerta Silva, 2008; Echeverri, 2008; Hernández Álvarez, 2000, 2013; Puerta Silva & Agudelo Calle, 2012; Yepes Luján, 2010).

La reforma del sistema de salud creó un modelo de salud mixto, regulado y financiado en parte por el Estado con la intermediación administrativa de empresas aseguradoras y prestadoras de servicios privadas y públicas. Desde entonces, se estructura sobre la administración del aseguramiento por parte de empresas quienes contratan con clínicas y hospitales los servicios de salud, organizados en planes de derechos. El Estado mantuvo para sí las funciones de financiación, regulación y de control. El esquema considera dos tipos básicos de regímenes de aseguramiento: el contributivo, dirigido a los empleados con capacidad de pago y, el subsidiado, para las personas sin capacidad de pago a quienes el Estado subsidia o presta directamente los servicios de salud.

Adicionalmente, por la misma época, las reformas democráticas en Colombia crearon y fortalecieron espacios y mecanismos de participación social y ciudadana en diversos campos de la vida pública y política. El sistema de salud, particularmente, acogió múltiples herramientas y promocionó la apertura de espacios de participación. Las novedades democráticas del sistema propiciaron, entonces, escenarios favorables para una relación ciudadano-Estado, a partir del reconocimiento de sus derechos y la focalización de subsidios para las personas sin capacidad de pago. Pero a pesar de la ampliación de los espacios y mecanismos de participación y una mayor inversión en el sistema de protección y seguridad social, la administración por parte de privados del aseguramiento y de la prestación de servicios ha resultado en la creación y consolidación de barreras de acceso, oportunidad y calidad a los servicios de salud que pueden ser leídas como vulneraciones y violaciones al derecho a la salud. En este contexto, los ciudadanos han debido no solamente adquirir habilidades burocráticas, sino que han tenido que emprender medidas jurídicas para lograr la protección de sus derechos a la salud. Se afirma que el sistema de salud se ha judicializado porque es a través de medidas jurídicas que los pacientes tienen acceso a la atención médica (Personería & Universidad de Antioquia, 2011).

En sus relatos, los televidentes se refieren a las barreras y constantes vulneraciones de su derecho a la salud, admitiendo que requieren mayor información sobre el funcionamiento del sistema y sobre sus derechos como ciudadanos. Es en este contexto particular, que la pregunta por el rol de la televisión en la formación ciudadana cobra importancia.

El concepto de ciudadanía que acogemos en este artículo la asume como un conjunto de prácticas, individuales y colectivas, que configuran modalidades de relacionamiento con el Estado, sus delegados, y con otros ciudadanos, las cuales se manifiestan en diversos ejercicios de derechos y deberes que incluyen su defensa, la participación y otras acciones que buscan el mantenimiento o el logro de la salud.

Dada la existencia de diversas combinaciones de prácticas (tanto individuales como colectivas y en los planos privados y públicos), se puede argumentar que no es una sola ciudadanía la que se produce en este contexto, sino múltiples. Comprender la ciudadanía más allá de la relación con el Estado y manifestada en plural, hace referencia a la producción de formas de convivencia que trascienden la pertenencia a un Estado. Esta perspectiva le da importancia por igual a una visión "horizontal" de la relación entre ciudadanos, como a una visión "vertical" tradicional de la ciudadanía que la entiende como la relación entre el individuo y el Estado (Kabeer, 2007). La ciudadanía resultaría, entonces, de un "contrato social" entre ciudadanos y entre estos y el Estado, que especifica "las formas de pertenencia y participación de los individuos en una comunidad" y que define su estatus sociopolítico, es decir, "los derechos y obligaciones [individuales y colectivos] que adquieren" (Haro, 2008). Según Haro, desde la década de 1980 el concepto de ciudadanía ha tomado fuerza, pues es el que coincide mejor con los nuevos espacios y mecanismos de participación popular en las aperturas democráticas, la descentralización de los poderes políticos y el reconocimiento de las pluralidades culturales, étnicas y de género, especialmente en los países latinoamericanos.

Estos procesos nacionales coinciden con la entrada de políticas económicas neoliberales, así como con el escepticismo de las poblaciones frente a los procesos electorales y la representatividad de sus líderes. En este contexto, la configuración de ciudadanías depende "de los términos en los que participan en esta vida colectiva y de las formas en las que pueden ejercer su capacidad de acción. Y cuando sólo pueden participar en términos de gran desigualdad, o cuando les es completamente negado el acceso, la ciudadanía se relaciona con sus intentos por desafiar estos procesos excluyentes y efectuar el cambio" (Kabeer, 2007). Nosotros identificamos la participación adversarial como parte de estos intentos (Dover & Puerta Silva, 2008; Puerta Silva & Agudelo Calle, 2012).

Múltiples ciudadanías, entonces, se producen debido a la correlación entre las políticas de reconocimiento y las reivindicaciones, intereses y acciones políticas que agencian las personas y que corresponden con valores como la justicia, el reconocimiento, la autodeterminación y la solidaridad (Kabeer, 2007); también con su posición en la sociedad, su identidad cultural, su situación espacial, su edad, su género y, en general, con las dinámicas globales que las afectan en su ser y estar de manera digna en el mundo. En el debate, particularmente en América Latina, se ha avanzado hacia la idea de las ciudadanías plurales y sobre el hecho de que no podemos asumir "*a singular, one-size-fits-all concept of citizenship and governance*" (De Munter & Salman, 2009, p. 436).

Para el caso exclusivo del ejercicio ciudadano con relación a la salud en Colombia, se ha identificado que la información y su comprensión son esenciales para la formación ciudadana y para el goce de salud y bienestar. Esta formación requiere de habilidades burocráticas y comunicativas para transitar por el sistema y ejercer sus derechos y deberes y ganar su lugar como ciudadanos. Aquí radica la pertinencia de estudiar la televisión y la comunicación de contenidos relacionados con la salud, para identificar cuál es la contribución de la televisión a la formación de ciudadanos. Adicionalmente a la potencia formativa que pueda tener la televisión, es necesario considerar que el televidente es un receptor activo, que deconstruye y le da significado a los contenidos televisivos de acuerdo a sus intereses y a unas mediaciones que resultan de su contexto experiencial y sociocultural (Agudelo Calle *et al.*, 2012).

De modo que los análisis de los contenidos televisivos o los formatos que se utilizan no son suficientes para entender la contribución de la televisión a la configuración de ciudadanías, pues solamente se concentran en la comprensión de la producción del mensaje y del contenido televisivo. No abordan el proceso de recepción y apropiación de contenidos que es tal vez más decisivo en la formación ciudadana y el que más recientemente ha sido estudiado (Martín-Barbero, 1987; Orozco Gómez, 1991, 1992). En efecto, los estudios de la recepción reconocen que hay un proceso cognitivo en la apropiación de los contenidos y mensajes, el cual puede caracterizarse a través de mediaciones que intervienen: culturales, individuales, situacionales e institucionales (Orozco Gómez, 1996). Los resultados presentados aquí permiten evidenciar en la televisión colombiana el aumento paulatino del tratamiento de temáticas relacionadas con salud y dan pistas sobre los factores que intervienen en su comprensión y traducción a prácticas ciudadanas por parte de los televidentes.

Métodos

La investigación que da lugar a este análisis utilizó varias técnicas dado su doble propósito: analizar los contenidos televisivos en salud y entender el proceso de apropiación de estos por parte de los televidentes, para determinar los aportes de este medio audiovisual a la formación ciudadana. Por ello se hizo uso del análisis textual para comprender los contenidos televisivos y técnicas etnográficas para entender la recepción del mensaje, esto es, la apropiación de los contenidos por parte de la audiencia.

Se seleccionaron cuatro programas de televisión que fueran representativos de lo que era emitido en ese momento en Colombia y que representaran a los canales privados y públicos y los nacionales y regionales. Así mismo que tuvieran el mayor contenido en salud, emisiones periódicas y que respondieran a diferentes tipos de géneros y formatos televisivos. Se eligieron en los dos canales privados nacionales, la sección de salud del noticiero de Caracol y el único *talk-show* dedicado exclusivamente a la salud, Doctor S.O.S. en RCN (réplica del Doctor OZ del grupo televisivo Fox). En el canal regional público TeleAntioquia, se seleccionó el magazine *Red Vital*; y en el canal local privado TeleVid, el magazine *Los Buenos Años*.

El análisis textual de los programas se realizó mediante la adaptación del "esquema de lectura" propuesto por Casetti y di Chio (1999). El esquema se enfocó en visibilizar: 1) definiciones y conceptos de salud (qué es salud y/o estar saludable), 2) estilos de vida y comportamientos saludables, 3) información sobre el sistema de salud, derechos y deberes (estructura y funcionamiento del sistema, acceso y uso de los servicios) y 4) temas que pudieran movilizar acciones políticas frente al sistema mismo o hacia su reforma.

Para abordar la recepción y apropiación de los contenidos, se hizo una exploración de tipo etnográfica por medio de entrevistas semiestructuradas a cuatro productores de televisión de los programas seleccionados y a treinta televidentes afiliados al régimen subsidiado de salud. Los televidentes fueron seleccionados de acuerdo a criterios de sexo (hombres y mujeres), nivel de escolaridad (primaria, bachillerato, universidad), rangos de edad (18 y 25; 26 y 40; 40 a 50 años y mayores de 50), relación con la televisión (ve con frecuencia televisión, combina la televisión con otros medios, la televisión es su medio exclusivo) y conocimiento sobre el sistema de salud (alto conocimiento y poco conocimiento); los dos últimos criterios se basaron en la autoclasificación de los participantes. Se buscó con estas entrevistas comprender el proceso de recepción y especialmente si los mensajes o contenidos se traducen en prácticas ciudadanas. Se realizaron, además, observaciones etnográficas a los televidentes mientras veían televisión y un grupo focal con ocho personas, basado en la discusión alrededor de los lenguajes, formatos, contenidos de cuatro segmentos televisivos y su traducción a prácticas ciudadanas.

Con base en la información obtenida a partir de las entrevistas se analizó el punto de vista de los entrevistados y se complementó con las observaciones hechas por los investigadores y por el contraste obtenido del grupo focal. A partir de esta indagación etnográfica, y con base en la información de otras investigaciones, se evidenció que la experiencia ciudadana en el sistema de salud está ligada a la vulneración constante de derechos. Por ello, fue necesario argumentar la existencia de las violencias del sistema para precisar mejor las particularidades de la formación ciudadana en salud en Colombia y el rol que juega la televisión.

Nuestro trabajo identifica el aporte que el contenido televisivo en salud tiene en la formación ciudadana, estudiando tanto el mensaje como la recepción e identificando si el televidente traduce lo que ve en la televisión en acciones que puedan ser consideradas como prácticas ciudadanas. Para ello, primero, se describirán los resultados de la investigación concernientes a los contenidos y formatos televisivos de los programas que versan sobre salud y, finalmente, se establecerán qué aspectos condicionan la recepción y su eventual traducción en ejercicios ciudadanos. En segundo lugar se hará la discusión sobre cómo las barreras y vulneraciones del goce efectivo del derecho a la salud devienen en violencias del sistema y cómo estas afectan la formación ciudadana en Colombia, identificando el aporte que la televisión podría tener.

1. Aportes de la televisión a la configuración de ciudadanías en salud

Se mencionó antes que asumimos que la formación ciudadana en salud se beneficia de la información que el ciudadano tenga sobre el sistema de salud, pero también de la apropiación de conceptos y prácticas que le permitan lograr el mantenimiento y/o recuperación de su salud y aportar a la construcción de condiciones adecuadas para el bienestar propio y colectivo. Es en el tema de la información y de la provisión de herramientas ciudadanas –instrumentos y recursos– donde pretendemos identificar el papel de la televisión.

Este medio masivo de comunicación es señalado por varios autores como vital en la formación y educación de los ciudadanos (Bustamante & Fernando, 2005; García Reina, 2004; Kubey & Csikszentmihalyi, 1990; Martín-Barbero, 2001; Pérez Tornero, 2005; Singhal & Rogers, 1999). Nuestros hallazgos revelan, por un lado, que la televisión aporta al conocimiento de enfermedades y de su prevención por la vía de la adopción de estilos y comportamientos saludables. Pero por otro lado, no aporta a la apropiación de conocimientos sobre derechos y, tampoco, sobre el funcionamiento del sistema; dicho de otro modo, no capacita a los televidentes en herramientas de participación. Para mostrar esto indicaremos, primero, los temas y formatos utilizados por las emisiones que fueron analizadas por este estudio, para luego señalar las mediaciones que condicionan o influencian la apropiación de contenidos y su transformación en acciones políticas.

1.1. Temas y formatos más utilizados para abordar la salud en televisión

Existe una demanda creciente por contenidos sobre salud en la televisión colombiana (Bonilla & Cadavid Bringe, 2004). En al menos los últimos cinco años han surgido *talk-shows*, magazines y secciones de noticieros pensados exclusivamente para abordar temáticas relacionadas con la salud física y mental. Los programas analizados cuentan con un amplio desarrollo de temáticas relacionadas con la definición de enfermedades y cómo remediarlas o prevenirlas mediante tratamientos médicos o estilos de vida, pero no abordan suficientemente, ni tienen una concepción clara, sobre temas que movilicen acciones políticas. Tampoco ofrecen una difusión amplia de información sobre las características y el funcionamiento del sistema que contribuya a un mejor conocimiento y reflexión de los ciudadanos sobre sus derechos y deberes y les permita transitar por el sistema ejerciendo o exigiendo un goce efectivo de sus derechos, beneficiándose de los servicios de salud y/o superando o confrontando las barreras/violencias del sistema.

Los programas de televisión estudiados ofrecen, en general, una amplia gama de contenidos que brindan información básica y universal sobre las enfermedades, sus síntomas, causas y hábitos para su prevención. Entre las temáticas desarrolladas se identificaron: el tabaquismo, los efectos de las olas de calor, el sonambulismo, el encaramiento del duelo, las principales lesiones de la columna vertebral, enfermedades de transmisión sexual, hábitos de higiene, la pubertad, la obesidad, beneficios de la actividad física, y la vejez, entre los más mencionados.

Desde el análisis textual de los programas –a partir del cual se le prestó especial atención a la intencionalidad y a los lenguajes empleados por los emisores a la hora de comunicar los contenidos–, y con base en entrevistas a sus productores, realizadores o responsables, se pudo establecer que predomina un marcado interés, con propósitos educativos, por suministrar contenidos que sean útiles al televidente-ciudadano para:

1. Conocer y reconocer las enfermedades: a través la divulgación de conocimientos básicos sobre ellas, representados en definiciones, ilustraciones y demostraciones.
2. Prevenir las enfermedades: por medio de la prevención y el estímulo de acciones que deriven en estilos de vida saludable; así mismo, haciendo llamados de atención sobre casos específicos de las enfermedades más comunes.
3. Buscar la asesoría de especialistas: ante anomalías que pudiera identificar relacionadas con su salud o la de sus familiares y amigos.

Los temas sobre las enfermedades y su prevención y, en menor medida, los estilos de vida saludables fueron más tratados en los magazines y el *talk-show*. Es evidente la ausencia de temas que impulsen movilizaciones políticas y la deficiencia de contenidos que ofrezcan al televidente-ciudadano un panorama esclarecedor y explicativo sobre el sistema de salud que rige en el país y, por ende, acerca de sus derechos y deberes como ciudadanos-usuarios. En las pocas ocasiones que estas temáticas se abordan, aparecen de manera tangencial durante el desarrollo de contenidos generales, pero con un despliegue parcial y que no supera una simple mención obligatoria, por estar vinculadas a la exposición de un tema principal. Es en la sección de salud de las noticias y en el *talk-show* analizado en donde se identificó la mención de temas que podrían inducir o promover movilizaciones y/o acciones políticas.

Esta carencia en los programas de televisión sobre salud es relevante si se tiene en cuenta las barreras de acceso a los servicios de salud que se han denunciado y el desconocimiento de los usuarios con respecto al funcionamiento del sistema de salud (Dover & Puerta Silva, 2008; Gañan Echavarría, 2013; Puerta Silva & Agudelo Calle, 2012; Puerta Silva, 2011), a sus características y a los roles que como sujetos de derechos y de deberes deberían asumir (Abadía Barrero & Oviedo Manrique, 2010; Hernández Álvarez, 2000).

Los temas sobre el sistema de salud, según la percepción de los productores y de los presentadores, son espinosos y requieren de conocimiento especializado. Ellos señalaron que serían irresponsables si trataran estos temas, porque los desconocen y porque requieren de una experticia de la que no gozan, además de un tratamiento amplio y profundo que demanda mucho más tiempo al aire. También indicaron que, en ocasiones, estos temas que denuncian fallas del sistema no son aceptados por los canales. No obstante, son enfáticos en afirmar que podrían ser muy exitosos, pues hay una necesidad sentida por los televidentes de contenidos que hablen del sistema de salud colombiano.

En lo que respecta a los formatos, encontramos que, en los cuatro canales seleccionados, los formatos de los programas con contenidos de salud son: noticieros, magazines y *talk-shows*. Los televidentes afirman que los dramatizados (como novelas y los seriados) y los *talk-shows* con ayudas didácticas utilizan lenguajes más apropiados para aprender temas de salud. Así mismo, algunos productores reconocen, como Ana Lucía Acosta del programa Red Vital, la potencia de los dramatizados y de los formatos que incluyen ayudas didácticas ilustrativas por ser más atractivos y transmitir mejor los contenidos, pero aceptaron que la selección del formato y de las estrategias didácticas depende directamente del costo de la producción del programa y, que por ello, sus programas tienen el formato de magazine, cuya producción es menos costosa.

Esto coincide con el registro de la literatura especializada de los formatos televisivos con mayor prelación: telenovelas, seriados, *reality shows*, magazines y noticieros y con la afirmación de que los formatos que logran conectar con las experiencias de los televidentes son más exitosos (Andaló, 2003; Beca & Salas, 2004; Christenson & Ivancin, 2006; Rideout, 2008). Se confirma que los enfoques de entretenimiento, dedicados al relato emotivo, no excluyen el enfoque educativo, y buscan formar, informar, trasmitir y transformar saberes. Al contrario, podrían coexistir los dos en un mismo lenguaje audiovisual, aunque obedece a la lógica del mercado del consumo y es determinado por las audiencias, pues el "[...] conjunto de sujetos sociales, activos e interactivos [...] entablan una relación situada con el referente mediático" (Concepción, 2008, p. 208). En síntesis, según estos análisis, si bien la televisión no se concibió para educar, su función no se agotaría en brindar entretenimiento, al contrario, podría posibilitar aprendizajes sin la necesidad de excluir la función de divertir.

La televisión colombiana, en conclusión, aporta información sobre las enfermedades, sus tratamientos y cómo prevenirlas mediante estilos de vida saludables desde el enfoque médico, lo cual deberá ser analizado con mayor profundidad en otras investigaciones con el fin de identificar los enfoques dominantes en el tratamiento de temáticas en salud. Los programas dedicados exclusivamente a la salud, excepto por el Doctor S.O.S., utilizan formatos poco atractivos que no logran detonar la vinculación emocional del televidente.

Dado que el hecho comunicativo no se agota en la transmisión del mensaje y bajo la evidencia que en la recepción televisiva juegan otros factores además del formato y contenido del mensaje, en la próxima sección avanzamos los resultados obtenidos respecto al proceso de la recepción.

1.2. Las mediaciones que afectan la apropiación de los contenidos televisivos

Entre los aspectos identificados por los televidentes como facilitadores de la apropiación de contenidos están: el lenguaje utilizado por los presentadores, la legitimidad y la credibilidad de los presentadores e invitados, y las ayudas didácticas para explicar e ilustrar. En este sentido, debemos hacer énfasis en que la credibilidad que le da el televidente al programa, resulta de otorgarle al personaje televisivo autoridad en el tema y cuando el mismo personaje se legitima con la utilización de indumentaria y un lenguaje técnico-médico. Esto coincide con las elecciones de los productores sobre los presentadores: la sección sobre salud de las noticias es liderada por una médica; el *talk-show* Doctor S.O.S. es presentado por un médico cirujano; y en los dos magazines del canal local y del regional actúan como presentadores un fisioterapeuta y una comunicadora social que se ha especializado en contenidos de salud.

Los televidentes afirmaron que en general no saben los horarios de los programas de salud. Solamente retienen de la parrilla de programación los horarios de noticieros y telenovelas o seriados. En el caso de los programas de salud, ellos reconocieron que se topaban con estos haciendo *zapping*, y continuaban viendo el programa si su temática estaba relacionada con sus intereses individuales o le concernía a alguna persona en su entorno cercano. Los televidentes entrevistados dijeron que consideran importante que haya programas de salud en la televisión pero no se declararon seguidores, excepto por el Doctor S.O.S. y Los Buenos Años en los televidentes mayores de 50 años.

Además de los formatos y lenguajes televisivos, las mediaciones cumplen un papel central en la apropiación de contenidos de salud. Martín-Barbero entiende las mediaciones como aquellas instancias culturales desde las cuales los significados y los sentidos son producidos y apropiados por la audiencia (1987, p. 203). En esta línea, Orozco (1992) define las mediaciones como el proceso mediante el cual se producen significados y sentidos a partir de la interacción con la televisión y con otras fuentes de significado sociocultural. Dicho de otro modo, "las mediaciones son una zona franca, donde cualquier significado es posible y donde no manda ninguna lógica cultural… son ese lugar desde donde es posible comprender la interacción entre el espacio de la producción y el de la recepción" (Martínez Terrero, 2006).

Nuestra búsqueda por entender el contexto sociocultural y la experiencia del televidente con el sistema de salud, coincide con la literatura que comprende que el proceso de recepción está influenciado por "la mente de la persona y el contexto socio-cultural en que se encuentra" (Martínez Terrero, 2006). Según Condry "la televisión influye en las acciones, los valores y las creencias de sus espectadores, pero no de la misma manera sobre todas las cosas" (1998, p. 82). La investigación entregó pistas para indagar por la interacción entre la televisión y televidente más allá del momento de estar frente a la pantalla, pues ella está condicionada desde antes y es también influenciada mientras transcurre y después que el televisor fue apagado (Orozco Gómez, 1996).

Dicho lo anterior, identificamos, gracias a las entrevistas, mediaciones emergentes (no identificadas en la literatura) y mediaciones ya descritas por otros autores. Como mediaciones emergentes encontramos: las violencias del sistema y las experiencias de los televidentes-ciudadanos con su salud y con el sistema de salud, y las prácticas de acción y movilización política en Colombia; y como mediaciones ya registradas: el entorno familiar, la escolaridad (Orozco Gómez, 1996), la edad y el género (Lacalle, 2012).

Confirmamos que en el entorno familiar, los conceptos y prácticas de salud y de estilos de vida, son transmitidos de generación en generación. Representan los primeros acercamientos al tema de la salud por parte de los televidentes y, en consecuencia, estos conocimientos previos condicionan de cierta manera la recepción de los contenidos televisivos.

En cuanto a la escolaridad, identificamos que determina, en buena medida, la credibilidad otorgada a la información de la televisión y la comprensión de los lenguajes televisivos. Quienes tenían un grado de escolaridad mayor, cuestionaban más lo expresado por los presentadores y los profesionales invitados en los programas. Además investigaban en otros medios, como internet, para contrastar y validar la información recibida. Se evidenció, en esta misma línea, que el lenguaje audiovisual que complementa la narrativa con ilustraciones gráficas (animaciones, videos, experimentaciones y objetos reales) es más comprensible para todo tipo de públicos, sobre todo para quienes tienen un grado de escolaridad menor.

Mediaciones como la edad y el género, por su parte, determinan las temáticas más atractivas para el televidente. Los más adultos, por ejemplo, se identificaron con temas como las principales lesiones de la columna vertebral y la vejez; mientras los más jóvenes se interesaron por tópicos como la nutrición, las enfermedades de transmisión sexual y el tabaquismo.

Por otro lado, entre las mediaciones emergentes, las violencias del sistema podrían estar contribuyendo a la forma en la que el ciudadano comprende y traduce los contenidos televisivos. En este sentido, las experiencias dramáticas con las barreras impuestas por el sistema se constituyen en una mediación, pues encontramos que a partir de estas experiencias, el televidente escoge o se conecta mejor con un tema tratado en la televisión, o se identifica con algo de lo que se esté abordando. Tales experiencias incluyen la relación con el médico y otro personal de salud, así como las experiencias administrativas dentro del sistema y el conocimiento adquirido gracias a la interacción con otros usuarios del sistema.

Ahora bien, la última de las mediaciones identificadas tiene que ver con las prácticas de acción y movilización política en Colombia, en especial en el sector de la salud. Se identificó que los entrevistados no participan de movilizaciones sociales o de acciones colectivas no, en parte por el desconocimiento de los mecanismos colectivos de acción. Al contrario, conocen y en algunos casos han utilizado mecanismos jurídicos como la Acción de Tutela[2]. Si bien los televidentes mencionaron que estaban de acuerdo con las movilizaciones cuyos propósitos son un sistema de salud más equitativo, reconocieron que nunca han participado en ellas. Se podría decir entonces que existe una tendencia a una movilización individual (a través de las tutelas) y no colectiva. Este elemento de la formación política de los televidentes puede considerarse una mediación en la medida en que no motiva ni favorece en el televidente la movilización política colectiva a pesar de que comparten las mismas experiencias de violencia por parte del sistema.

En los siguientes apartes hacemos la discusión sobre la importancia de considerar en el análisis del aporte que los medios pueden tener en la formación ciudadana, el contexto general de lo que hemos denominado aquí las violencias del sistema. Las prácticas ciudadanas, en este contexto, se concentran en la defensa, protección y restitución de derechos, más que en la participación política en el desempeño del sistema o en el debate público sobre los derechos en salud. Podría concluirse entonces, que la televisión aporta parcialmente con información sobre las enfermedades y su tratamiento, pero no contribuye al debate público sobre el sistema de salud ni a la formación de habilidades específicas ciudadanas para hacer frente a las violencias de salud sufridas por los televidentes.

1.3. Violencias del sistema de salud colombiano

En el análisis de la relación entre salud y violencia, esta última se define por la Organización Mundial de la Salud como "El uso deliberado de la fuerza física o el poder, ya sea en grado de amenaza o efectivo, contra uno mismo, otra persona o un grupo o comunidad, que cause o tenga muchas probabilidades de causar lesiones, muerte, daños psicológicos, trastornos del desarrollo o privaciones" (Organización Mundial de la Salud, 2002, p. 5). En general, es abordada como un factor externo que

---

2 La acción de tutela es un procedimiento amparado por la Constitución Política de Colombia para demandar la protección inmediata de los derechos constitucionales fundamentales, cuando estos sean vulnerados por la acción u omisión de cualquier autoridad pública o de particulares.

tiene implicaciones directas sobre la salud o como un comportamiento a prevenir. Los sistemas de salud y los profesionales se han preocupado por entender la diversidad de las formas en las que se ejerce violencia, los efectos de ella sobre la salud, sus causas y las posibles respuestas, las alternativas o políticas para prevenirla o tratarla.

La violencia en su relación con la salud se clasifica en tres categorías: violencia dirigida contra uno mismo, violencia interpersonal y violencia colectiva (Organización Mundial de la Salud, 2002). Esta última, puede ser social, política y económica y se puede manifestar, en general, como la violación de derechos humanos y, en particular, puede tomar la forma de privación o desatención.

Las barreras de acceso, oportunidad y calidad en los servicios de salud pueden ser leídas como vulneraciones y violaciones al derecho a la salud, por la intermediación financiera y las prácticas de racionalidad del costo, propias del sector privado. Han sido documentadas tanto estas barreras, como la decadencia en la calidad técnica y humana de los servicios. También se han evidenciado los indicadores negativos en la prevención de enfermedades ya erradicadas o de maternas e infantes y en la promoción de estilos de vida saludables. Otros de los factores que generan violencia y que afectan directamente a los ciudadanos en su goce efectivo de sus derechos, y a la sociedad en su conjunto por estar enferma y desprotegida, son: la tendencia a una medicalización en el sistema, esto es, servicios dirigidos exclusivamente a la gestión de la enfermedad; la burocratización y la judicialización del sistema; la corrupción y el usufructo de dineros públicos por parte de aseguradores y prestadores de servicios; el traslado de costos a los pacientes (Abadía Barrero & Oviedo Manrique, 2010; Alvarado Romero, Pinilla Alarcón, Rodríguez Colmenares, & Castro Suárez, 2010; Yepes Luján, 2010).

> Servicios negados. Atención negligente. Pacientes que mueren en los pasillos esperando una autorización de su EPS [Entidad Promotora de Salud]. Hospitales atestados. En este país nos acostumbramos a padecer un sistema que funciona mal. Las cifras así lo evidencian: son cada vez más las quejas, los reclamos y las tutelas que buscan un servicio digno, que salve vidas. Mientras tanto, el Gobierno, al declarar la salud como un derecho fundamental, busca salidas a la crisis[3].

Asumimos entonces que la privación, la desatención y otras omisiones provocadas por las transformaciones administrativas y económicas del sistema de salud colombiano, tienen como consecuencia la violación

---

3 El Tiempo. Reportaje de datos sobre las tutelas y quejas en el sistema de salud de Colombia, disponible en: <http://www.eltiempo.com/multimedia/infografias/salud/14436324>, consultado el 24 de septiembre de 2014.

reiterativa del derecho al acceso a los servicios de salud. Y en ese sentido, dicha violación reiterativa de los derechos por parte del sistema revela el ejercicio de la violencia infringida por el Estado y sus delegados (privados) en términos de privaciones o desatención (Organización Mundial de la Salud, 2002). Son violencias porque tienen consecuencias para la salud y la vida, como son: la prolongación del sufrimiento, cuando se causa angustia emocional por el tiempo que las personas y sus allegados deben esperar para ser atendidas y recibir tratamiento; las complicaciones médicas del estado de salud por esta espera; los daños o la discapacidad permanentes o hasta la muerte (Abadía Barrero & Oviedo Manrique, 2010).

Debido a las barreras, desatenciones, problemas de calidad y oportunidad, trato inhumano, dilatación u omisión de la atención y el cuidado, medicalización del sistema, etc., se están provocando daños, lesiones e incluso muerte a los ciudadanos. Sin embargo, aunque se denuncian las muertes ocasionadas por la estructura mixta del sistema, principalmente por la participación de las aseguradoras privadas, no hay datos exactos (Gañan Echavarría, 2013). Solo contamos con los datos de las acciones de tutelas en salud, que revelan desde el punto de vista del ciudadano, las violencias de las cuales es objeto (Colprensa, 2012; Personería de Bogotá, 2008; Vélez, 2005; Vélez-Arango, Realpe-Delgado, Gonzaga-Valencia, & Castro-Castro, 2007). Para el año 2013 se interpusieron 115.147 acciones de tutela en salud. Según la Defensoría del Pueblo, las causas más frecuentes se refieren a la negación o retrasos en la autorización de tratamientos 26%, medicamentos 15%, citas médicas 11%, cirugías 9% y prótesis e insumos médicos 9% (El Tiempo, 26 de agosto de 2014).

2. La configuración de ciudadanías en el contexto de las violencias del sistema

El senador Jorge Enrique Robledo ha asegurado que la Ley 100 –Ley que transformó el sistema de salud colombiano– "mata más colombianos que todas las violencias que nos martirizan"[4]. En su recorrido por los "itinerarios burocráticos" (Abadía Barrero & Oviedo Manrique, 2010) o en su confrontación con las violencias del sistema, los ciudadanos-pacientes cada vez más recurren a la exigibilidad jurídica. En este contexto, se ven en la necesidad de adquirir información y de formar capacidades para transitar por el sistema.

> El sistema de salud de Colombia ha creado nuevos tipos de itinerarios en los que la atención en salud no depende de las necesidades de las personas o de la valoración médica, pero sí del cumplimiento exitoso de las normas

4 La salud por el Senador Jorge Enrique Robledo, disponible en: <http://www.youtube.com/watch?v=iMQcuQ8VXl0>, consultado el 24 de septiembre de 2014.

administrativas del sistema y los costos financieros (nuevas barreras de acceso y calidad). Cuando es imposible recibir la atención necesaria, el acceso depende de *conocer y realizar acciones legales*, cuyos resultados se basan no sólo en las interpretaciones de los jueces, sino también en el cumplimiento de los fallos por parte de las instituciones (Abadía Barrero & Oviedo Manrique, 2010).

Dicho de otra manera, los ciudadanos hoy deben movilizarse para restituir el derecho fundamental de la salud, de modo que esta deje de ser "una mercancía" y "no [haya] más negocio con el sufrimiento" (Hernández Álvarez, 2013). La transformación del sistema, dirigido hacia una lógica mixta de mercado, al tiempo que supone ser un servicio público, complejiza los roles que cumple el paciente. Por un lado, su rol como ciudadano, sujeto de derechos y, por otro, su rol como cliente-usuario de unos servicios prestados y regulados por empresas aseguradoras y por hospitales y clínicas. Dicha complejidad se debe entonces a la coexistencia de estas dos lógicas o estructuras –la privada, dirigida al mercado y, la pública, dirigida a la garantía de derechos en salud– (Agudelo Calle, 2009). En este contexto, pacientes y agentes de salud se ven obligados a conocer el sistema para transitar en él y conocer sus roles y responsabilidades, combinando capacidades administrativas, especializadas en salud, y habilidades comunicativas, en medio de eventos dramáticos cuando sus derechos son vulnerados.

Cuando presumimos que se configuran múltiples ciudadanías en salud, estamos reconociendo las agencias y experiencias de los sujetos y, en este sentido, la producción de ejercicios ciudadanos y de ciudadanías que se presentan como respuestas, acomodamientos o, incluso, alternativas a las prácticas estatales (Gupta, 1995). En resumen, reconocemos que la configuración de ciudadanías resulta de un diálogo entre 1) saberes y prácticas cotidianas frente a la salud y a la enfermedad; 2) representaciones sobre el Estado y el posicionamiento de individuos y colectivos con respecto a él; 3) los conocimientos y las experiencias con el sistema de salud, con los derechos y deberes y con la participación institucional; y 4) la información y las capacidades para transitar y accionar políticamente en el sistema, entre otros factores.

Creemos que es necesaria esta visión más amplia de la ciudadanía, integrando en ella los comportamientos y actos cotidianos que inciden en la salud individual y colectiva, así como el accionar político en múltiples modalidades. Por ello, asumimos que las ciudadanías en salud implican un ejercicio informado de toma de decisiones que redundan en la salud individual y colectiva, pero, además, en el contexto de un sistema de salud que ejerce violencias contra sus ciudadanos, estas ciudadanías involucran la apropiación de una serie de herramientas de protección y restauración de derechos vulnerados.

Hemos identificado que las ciudadanías no solo son múltiples sino que se están configurando parcialmente. Lo más visible desde nuestra perspectiva es la activación de prácticas de participación que se activan de manera situacional o coyuntural (Puerta Silva & Agudelo Calle, 2012). Las personas no asumen que su ciudadanía se ejerce cada día en la relación intrínseca en el ámbito público con otros ciudadanos y con el Estado o sus delegados, sino que solo son conscientes de ella cuando requieren accionar mecanismos de participación, como la tutela, en ocasión de situaciones dramáticas, producto de las diversas violencias del sistema, lo cual consolida ciudadanías de tipo adversarial, esto es prácticas radicales de protección o demanda de derechos por vías judiciales y no ejercicios ciudadanos construidos a partir del goce efectivo del derecho a la salud.

Con la medicalización del sistema de salud, sus violencias y la carencia de educación política, la mayoría de las personas no tienen ni los recursos informativos ni las capacidades para configurar ciudadanías integrales (que incluyan tanto los comportamientos saludables como la participación propositiva en el sistema), sino que deben concentrarse en las prácticas adversariales –acción de tutela, derecho de petición, queja–, es decir, en los mecanismos de exigibilidad jurídica del derecho. La Defensoría del Pueblo de Colombia señala que la Acción de Tutela se convirtió en un "prerrequisito de trámite" para tener acceso a los servicios. Según ellos, cada dos minutos se instaura una acción de exigibilidad jurídica (Colprensa, 2012). Ratificamos entonces, que el sistema se ha judicializado, debido a que una buena parte de las acciones de participación ciudadana son de exigibilidad jurídica, esto es, pretenden denunciar una vulneración y/o solicitar el goce efectivo de un derecho.

Las ciudadanías en tanto prácticas sociales, políticas y culturales, deberían, en su condición ideal, representar las necesidades e intereses de las personas y los colectivos adecuados y efectivamente, de manera que pudieran disputarse los fallos, barreras y desigualdades en los servicios, esto es, la desatención y la privación que, argumentamos, expresan las violencias del sistema de salud en Colombia. Hoy, la alta incidencia de acciones de tutela y otros mecanismos jurídicos, así como la activación de acciones y estrategias no institucionales para acceder a los servicios o sobreponerse a obstáculos en la prestación de los mismos, demuestran que las ciudadanías adversariales dominan el campo de relacionamiento con el sistema de salud (Dover & Puerta Silva, 2008; Echeverri, 2008; Gañan Echavarría, 2013; Puerta Silva & Agudelo Calle, 2012).

En este sentido, las ciudadanías se están configurando en confrontación con el Estado y no en una relación complementaria o dialógica en clave democrática. Por ello, se activan los ejercicios ciudadanos ante sentimientos de indignación, por la violación de los valores que rigen el goce de los derechos ciudadanos. Las violencias del sistema están promoviendo, desde nuestra perspectiva, la configuración de ciudadanos adversariales, en confrontación con el Estado. Pero además de fortalecer modalidades adversariales de participación, las violencias del sistema, han promovido también la activación estrategias sociales que los pacientes desarrollan gracias a su experiencia y que están por fuera de los canales institucionales para resolver asuntos dentro del sistema de salud (Dover & Puerta Silva, 2008; Puerta Silva & Agudelo Calle, 2012). La televisión, por su parte, no está aportando al debate público sobre la estructura del sistema de salud y los efectos que han sido evidenciados durante los últimos 10 años y que hemos denominado aquí violencias del sistema. Y, ante la ausencia de contenidos que puedan movilizar la acción política para demandar la reforma del sistema, tampoco está contribuyendo a que los televidentes tengan los elementos para ejercer la dimensión de la ciudadanía en salud, que consiste en proteger su derecho a la salud.

Conclusiones

Un sistema de salud que propició una apertura democrática para los ciudadanos colombianos y amplió la cobertura del aseguramiento, se ha convertido al mismo tiempo en un causante de violencias. Las múltiples vulneraciones a los derechos por las barreras al acceso, la oportunidad y la calidad de los servicios, el traslado de costos a los pacientes, el enriquecimiento de empresas privadas con dineros públicos, el aumento de la prevalencia de enfermedades prevenibles, el sufrimiento y la muerte, son algunas manifestaciones de dichas violencias.

Es en este contexto en el que se han configurado las ciudadanías y se han formado los usuarios del sistema como ciudadanos. Por ello, a pesar de los múltiples espacios y mecanismos de participación que la normatividad ha generado y que se dirigen hacia la contribución de los ciudadanos a un mejor desempeño del sistema, los mecanismos más activados han sido adversariales. Son acciones de exigibilidad jurídica que buscan proteger o recuperar derechos en salud, particularmente a través de la Acción de Tutela.

Hemos argumentado que para la formación de ciudadanías en salud son fundamentales la información y el conocimiento sobre la salud, la enfermedad, los comportamientos saludables, el sistema y los derechos, además de las habilidades o capacidades para transitar por el sistema y ejercer los derechos. Nos preguntamos si la televisión colombiana contribuía en este aspecto. Se evidenció que han aumentado las emisiones que contemplan temas relacionados con la salud. Confirmamos que la televisión tiene un gran potencial para informar y formar a los televidentes en asuntos relativos a su salud. En efecto, la televisión colombiana está aportando información sobre enfermedades y algunas estrategias para prevenirlas o atenderlas. Los formatos televisivos más utilizados son los magazine y *talk-shows*, mientras que los dramatizados son los menos utilizados a pesar de que se reconoce su potencia educativa en salud (Arroyave, 2008). Los costos de producción parecen estar al origen de la selección del formato, por lo menos en el caso de los canales locales y regionales. Todos reconocen que el formato del *talk-show*, especialmente por los recursos visuales y gráficos complementarios, es el más exitoso al igual que los dramatizados.

Establecimos que en los programas analizados había una prelación por contenidos relativos a estilos de vida saludables y a conceptos de salud biomédicos, con un enfoque desde la enfermedad y su prevención, pero una ausencia generalizada de temas relativos al sistema de salud colombiano y a los derechos y deberes. No logramos identificar contenidos que motivaran o incitaran acciones o movilizaciones políticas. Con relación a este último tema, los productores reconocieron su importancia pero no lo abordan por criterios de los canales o porque requieren un tratamiento especializado.

En cuanto a la apropiación de contenidos, confirmamos que esta depende de las mediaciones del contexto sociocultural y de las experiencias del televidente, así como del contexto sociopolítico que ha definido la estructura y funcionamiento del sistema de salud. No fue evidente que los contenidos emitidos por los programas analizados les sirvieran a los televidentes para su acción política. En cambio, si parecieron ser traducidos en acciones, cuando fueron incorporados a las prácticas cotidianas de prevención de la enfermedad. Los programas analizados promueven un enfoque individual de la salud. Invisibilizaron la dimensión colectiva, es decir, la salud como un propósito social común y no solamente personal. En este sentido, la televisión no aporta a explicitar la dimensión colectiva de las ciudadanías y no son expuestas las posibilidades de participación ciudadana en su desarrollo.

Identificamos que algunas de las interpretaciones de los contenidos televisivos se vinculan con las reacciones emocionales que generan. Hallamos en otras investigaciones que un motivante importante de la participación ciudadana responde precisamente a las emociones detonadas por eventos dramáticos o por experiencias violentas en el sistema. Entender las violencias del sistema de salud, permite relacionar su carácter dramático con los motivantes de la participación y el escenario favorable para la activación de modalidades de participación adversarial.

¿Cómo promover ciudadanías corresponsables, solidarias, relacionales, reconociendo la pluralidad de configuraciones posibles para diferentes agentes sociales en múltiples contextos y situaciones? ¿Cómo formar ciudadanos en salud que cotidianamente agencien comportamientos que mantienen o recuperan la salud y que busquen y se apropien de información y la comprenden suficientemente para tomar decisiones que propendan por su salud y la de su entorno más cercano?

Hemos argumentado que la formación ciudadana exige el conocimiento para realizar trámites burocráticos que permitan transitar adecuadamente por el sistema de salud y que mantengan la motivación de actuar e incidir de manera constructiva y propositiva en la sociedad, con acciones y movilizaciones políticas que permitan el goce efectivo de derechos en salud y el cumplimiento de los deberes y la participación en el mejoramiento del desempeño del sistema de salud. Es a partir de estas cuestiones, que creemos pertinente seguir indagando por el aporte de la televisión.

Por un lado, el tratamiento televisivo limitado a las enfermedades y a su prevención, no aporta desde nuestra perspectiva al debate sobre el sistema de salud, ni sobre las violencias que ejerce sobre los ciudadanos. No contribuye con contenidos que puedan ser traducidos en acciones políticas. Por otro lado, la experiencia y el conocimiento sobre salud de los televidentes han estado atravesados por las violencias del sistema de salud, y en esta perspectiva, su lectura y apropiación de los mensajes televisivos, se detiene en el aprovechamiento parcial de la información sobre las enfermedades y su prevención, pero no se beneficia de información para su accionar político. Estos hallazgos permiten avanzar en la comprensión de la influencia de los medios masivos en la formación política y de comportamientos saludables, aspectos centrales en la configuración de ciudadanías en salud.

Referencias

Abadía Barrero, C. E., & Oviedo Manrique, D. G. (2010). "Itinerarios burocráticos de la salud en Colombia: la burocracia neoliberal, su estado y la ciudadanía en salud". *Revista Gerencia Y Políticas de Salud*, 9(18), 86-102.

Agudelo Calderón, C. A., Cardona Botero, J., Ortega Bolaños, J., & Robledo Martínez, R. (2011). "Sistema de salud en Colombia : 20 años de logros y problemas". *Ciência & Saúde Colectiva,* 16(6), 2817-2828. doi:10.1590/S1413-81232011000600020.

Agudelo Calle, A. (2009). *La relación médico-paciente en la Unidad Hospitalaria Jaime Tobón Arbeláez: una mirada antropológica*. Medellín: Universidad de Antioquia.

Agudelo Calle, A., Puerta Silva, C., Hernández Marín, A., Gómez Jaramillo, Y., Hernández, A., & Rendón, M. (2012). *Proyecto de investigación: La televisión y sus aportes a la construcción de ciudadanías en salud.* Medellín: Universidad de Antioquia Iner y Codi.

Alvarado Romero, H. R., Pinilla Alarcón, M., Rodríguez Colmenares, S. M., & Castro Suárez, E. M. (2010). "Sistema General de Seguridad Social en Salud y decretos de emergencia social: negación del cuidado de la salud". *Revista Gerencia Y Políticas de Salud*, 9(18), 53-60.

Andaló, P. (2003). "Lágrimas, amores, traiciones y mensajes de salud". *Perspectivas de Salud*, 8(2), 8-13. Retrieved from <http://cidbimena.desastres.hn/docum/ops/Revistas/persp17span.pdf>.

Arroyave, J. (2008). "Testing the effectiveness of an entertainment-education health-focused soap opera: exposure and post-discussion in colombian young adults". *Investigación Y Desarrollo*, 16(2), 232-261.

Beca, J. P., & Salas, S. I. (2004). "Medicina en televisión: ¿un problema ético?". *Revista Médica de Chile*, 132, 881-885. doi:10.4067/S0034-98872004000700015.

Bonilla, J. I., & Cadavid Bringe, A. (2004). "Los ciudadanos. Audiencias, agendas y demandas de información". In J. I. Bonilla & A. Cadavid Bringe (Eds.), *¿Qué es noticia? Agendas, periodistas y ciudadanos* (pp. 75-94). Bogota: Pontificia Universidad Javeriana y Fundacion Konrad Adenauer.

Bustamante, B., & Fernando, A. (2005). "Televisión de calidad y participación ciudadana". *Comunicar*, 25.

Casetti, F., & di Chio, F. (1999). *Análisis de la televisión. Instrumentos, métodos y prácticas de investigación*. Barcelona: Paidós.

Christenson, P., & Ivancin, M. (2006). *The "Reality" of Health: Reality Television and the Public Health. Discussion paper*. Menlo Park: Henry J. Kaiser Family Foundation. Retrieved from <http://kaiserfamilyfoundation.files.wordpress.com/2013/01/7567.pdf>.

Colprensa (2012, December 20). "Por reclamos en servicios de salud, cada dos minutos es puesta una tutela". *El Colombiano*. Medellín. Retrieved from <http://www.elcolombiano.com/BancoConocimiento/S/salud_en_colombia_cada_dos_minutos_se_pone_una_tutela/salud_en_colombia_cada_dos_minutos_se_pone_una_tutela.asp>.

COMFER, C. F. de R. (2007). *Investigación sobre tv y salud. Estudio sobre los mensajes de salud en la televisión abierta*. Buenos Aires: COMFER Comité Federal de Radiodifusión. Retrieved from www.afsca.gob.ar/cwp-content/uploads/2009/estudio-tv-y-salud-parte-11.doc.

Comisión Nacional de Televisión (2010). *Módulo de Televisión EGM Estudio General de Medios III-2010 (Tercer Ola 2010)*. Bogotá: Comisión Nacional de Televisión.

Condry, J. (1998). "Ladrona de tiempo, criada infiel". In *La televisión es mala maestra*. México, D.F.: Fondo de cultura económica.

DANE (2011). *Encuesta de cultura 2010*. Bogotá: DANE.

De Munter, K., & Salman, T. (2009). "Extending Political Participation and Citizenship: Pluricultural Civil Practices in Contemporary Bolivia". *The Journal of Latin American and Caribbean Anthropology*, 14(2), 432-456. doi:10.1111/j.1935-4940.2009.01055.x.

Dover, R. V., & Puerta Silva, C. (2008). *El derecho a la salud: la participación en el régimen subsidiado*. Medellín: Universidad de Antioquia INER y Colciencias.

Echeverri, O. (2008). "Mercantilización salud para el desarrollo: el caso de Colombia". *Revista Panamericana de Salud Pública*, 24(5), 210-216. Retrieved from <http://www.scielosp.org/pdf/rpsp/v24n3/a08v24n3>.

Gañan Echavarría, J. L. (2013). *Los muertos de la Ley 100. Prevalencia de la libertad económica sobre el derecho fundamental a la salud: una razón de su ineficiencia. Caso del Plan Obligatorio de Salud en el Régimen Contributivo*. Medellín: Universidad de Antioquia, Facultad de Derecho y Ciencias Políticas.

García Reina, L. (2004). "Juventud y medios de comunicación. La televisión y los jóvenes: Aproximación estructural a la programación y los mensajes". *Ámbitos*, (12), 115-129. Retrieved from <http://grupo.us.es/grehcco/ambitos11-12/reina.pdf>.

Gupta, A. (1995). "Blurred Boundaries: The Discourse of Corruption, the Culture of Politics, and the Imagined State". *American Ethnologist*, 22(2), 375-402.

Haro, J. A. (2008). "Construcción de ciudadanía en salud". In *XV Curso-Taller OPS/OMS-CIESS Legislación de salud: marco regulatorio para la extensión de la protección social en salud*. México, D.F. Retrieved from <http://www.slideshare.net/JESUSARMANDOHARO/ciudadana-en-salud>.

Hernández Álvarez, M. (2000). "El Derecho a la salud en Colombia: obstáculos estructurales para su realización". *Revista de Salud Pública*, (2), 121-144.

Hernández Álvarez, M. (2013). "Momento crítico para la salud en Colombia". *Seminario Virtual Caja de Herramientas*, (339). Retrieved from <http://viva.org.co/cajavirtual/svc0339/articulo03.html>.

Kabeer, N. (2007). "En busca de una ciudadanía incluyente: sus significados y expresiones en un mundo interconectado". In N. Kabeer (Ed.), *Ciudadanía incluyente: significados y expresiones* (pp. 7-33). México, D.F.: Universidad Nacional Autónoma de México.

Kubey, R. W., & Csikszentmihalyi, M. (1990). *Television and the Quality of Life: How Viewing Shapes Everyday Experience*. Hillsdale: Routledge.

Lacalle, C. (2012). "Género y edad en la recepción de la ficción televisiva". *Comunicar, Revista Científica de Educomunicación*, XX(39), 111-118. doi:10.3916/C39-2012-03-01.

Martín-Barbero, J. (1987). *De los medios a las mediaciones*. México: Editorial Gustavo Gili, S.A.

Martín-Barbero, J. (2001). "Televisión pública, televisión cultural: entre la renovación y la invención". In *Televisión pública: del consumidor al ciudadano* (pp. 1-21). Bogotá: FES/Promefes.

Martínez Terrero, J. (2006). "Teorías de comunicación". *Ágora RIIAL*. Ciudad Guyana: Universidad Católica Andres Bello, Núcleo Guayana. Retrieved March 9, 2013, from <http://www.riial.org/espacios/teoriacom/teoriacom_docbase.pdf>.

Molina, G., & Ramírez, A. (2013). "Conflicto de valores en el sistema de salud de Colombia: entre la economía de mercado y la normativa constitucional, 2007-2009". *Revista Panamericana de Salud Pública*, 33(4), 287-293. doi:10.1590/S1020-49892013000400008.

Obregón Gálvez, R. R., Vega Casanova, M. J., & Lafaurie Molina, A. (2003). "Medios y salud pública: la voz de los adolescentes". *Investigación Y Desarrollo*, 11(2), 210-237.

Organización Mundial de la Salud (2002). *Informe mundial sobre la violencia y la salud. Resumen*. Washington D.C.: Organización Mundial de la Salud.

Orozco Gómez, G. (1991). *Recepción Televisiva: Tres aproximaciones y una razón para su estudio. Cuadernos de Comunicación y Prácticas Sociales*. México, D.F.: Universidad Iberoamericana.

Orozco Gómez, G. (1992). "La investigación de la recepción y la educación para los medios: hacia una articulación pedagógica de las mediaciones en el proceso comunicativo". In M. Miranda (Ed.), *Educación para la Comunicación: Manual Latinoamericano*. Santiago de Chile: Ceneca.

Orozco Gómez, G. (1996). *Televisión y audiencias. Un enfoque cualitativo*. (E. de la T. y U. Iberoamericana, Ed.). Madrid.

Parra, L. N., & Oliva, G. (2004). "Análisis de la oferta televisiva en promoción para la salud y en salud sexual para los adolescentes". *Revista de Ciencias Sociales*, X, 112-130. Retrieved from <http://www.redalyc.org/articulo.oa?id=28010109>.

Pérez Tornero, J. M. (2005). Los ciudadanos y la televisión. *Analisi*, 32, 251-256.

Personería de Bogotá (2008). *Acción de tutela*. Bogotá. Retrieved from <http://www.personeriabogota.gov.co/?idcategoria=407>.

Personería, & Universidad de Antioquia (2011). *Experiencias y lecciones de un observatorio de salud en Colombia*. (D. *et al*. Guerra, Ed.). Medellín: Ediciones La Carreta.

Puerta Silva, C. (2011). "Prácticas de reconocimiento y visiones del Estado en el régimen subsidiado de salud colombiano". In M. Chaves (Ed.), *La multiculturalidad estatalizada. Indígenas, Afrodescendientes y configuraciones regionales de Estado* (pp. 211-223). Bogotá: Instituto Colombiano de Antropología e Historia.

Puerta Silva, C., & Agudelo Calle, A. (2012). "Modalidades de interacción en el Régimen Subsidiado de Salud en Medellín: Participación y estrategias sociales". *Hacia La Promoción de La Salud*, 17(1), 13-28. Retrieved from <http://promocionsalud.ucaldas.edu.co/downloads/Revista17(1)_1.pdf>.

Rideout, V. (2008). *Television as a health educator: a case study of Grey's Anatomy. Victoria* (Kaiser Fam.). California: Henry J. Kaiser Family Foundation. Retrieved from <http://kaiserfamilyfoundation.files.wordpress.com/2013/01/7803.pdf>.

Silverstone, R. (2007). *La moral de los medios de comunicación. Sobre el nacimiento de la polis de los medios*. Buenos Aires: Amorrortu.

Singhal, A., & Rogers, E. M. (1999). *Entertainment-Education: A Communication Strategy for Social Change*. Mahwah: Erlbaum. Retrieved from <http://books.google.com.co/books?id=FWNPUpuLav4C>.

Vargas J., & Molina G. (2009). "Acceso a los servicios de salud en seis ciudades de Colombia: limitaciones y consecuencias". *Rev Fac Nac Salud Pública*, 27(2), 121-130.

Vélez, A. L. (2005). "La acción de tutela: ¿un mecanismo de protección del derecho a la salud y un proceso alterno para acceder a servicios de salud?" *Colombia Médica*, 36(3), 199-208.

Vélez-Arango, A. L., Realpe-Delgado, C., Gonzaga-Valencia, J., & Castro-Castro, A. P. (2007). "Acción de Tutela, Acceso y Protección del Derecho a la Salud en Manizales, Colombia". *Rev Salud Publica (Bogota)*, 9(2), 297-307.

Yepes, F., & Ramírez, M. (2007). *Pérdidas y ganancias de las reformas en salud. Universidad, Ciencia y Desarrollo* (Vol. II). Bogotá: Universidad del Rosario. Retrieved from <http://www.urosario.edu.co/urosario_files/05/0504bf63-aa7c-4483-81f6-b9f8f523dd92.pdf>.

Yepes Luján, F. J. (2010). "Ni emergencia, ni salud de la población". *Revista Gerencia Y Políticas de Salud*, 9(18), 30-40. Retrieved from <http://www.redalyc.org/pdf/545/54520413003.pdf>.

Part 2

Violence and the City

Ciudad, conflicto y paz en Colombia[1]

Carlos Mario Perea

En el proceso de paz iniciado a finales de 2012 en la Habana, ¿dónde está la ciudad?[2]. En Colombia, nación donde las tres cuartas partes de la población habitan la urbe, la pregunta tiene toda pertinencia. La agenda de cinco puntos pactada entre el gobierno nacional y las FARC –reforma agraria, participación política, narcotráfico, víctimas y reintegración civil de los combatientes–, no contempla de ningún modo el conflicto de la ciudad. La paz se entiende al modo de la remoción de causas estructurales que están al origen del conflicto armado, ninguna de las cuales apunta a la superación de la violencia urbana. Entre la ciudad y la paz hay entonces un abismo que abre un interrogante sobre la visión que ha dominado la conciencia pública sobre el conflicto y las maneras de tramitarlo. El contrapunteo entre las nociones de conflicto armado y conflicto violento, develando el lugar que la ciudad ocupa frente al uno y el otro, servirá como hilo conductor de las presentes páginas. Nos guía una afirmación esencial: la paz no es sólo la finalización del conflicto armado, es, más allá, la supresión del conflicto violento.

De partida, la importancia capital de la ciudad no se circunscribe a la condición de país urbano. Entre los años de 1980 y 2010 se cometieron en Colombia 652.356 homicidios, de los cuales 364.716 –es decir el 56%–, tuvieron lugar en las ciudades (DANE, 1980-2010). Tal cantidad de homicidios deja fuera de duda la importancia del contexto urbano en la búsqueda de la paz; finalmente durante las últimas tres décadas, las más

---

1 Una versión inicial del artículo apareció publicado en *Análisis Político* bajo el título de "Resituar la Ciudad. Conflicto violento y paz". No. 77, enero-abril de 2013.

2 El artículo hace parte de la investigación comparada de los conflictos violentos de Medellín, Bogotá, Río de Janeiro y Ciudad Juárez realizado por la Corporación Región, la Universidad Federal de Río de Janeiro, el Colegio de Chihuahua y el Instituto de Estudios Políticos y Relaciones Internacionales (IEPRI) de la Universidad Nacional de Colombia. La financiación proviene de Colciencias y del International Development Research Centre (IDRC) de Canadá. Su desarrollo hubiera sido imposible sin el decidido aporte de Andrés Rincón, Sayra Rodríguez, Nancy Guzmán, Gloria Bermúdez, María Eugenia González y Ricardo Moreno, el grupo de investigación "Ciudad y conflicto" del IEPRI.

convulsas de la historia moderna del país, algo más de la mitad de las muertes letales fueron producidas en la calle de la ciudad[3].

Sin embargo, pese a su enorme contribución, la ciudad mantiene un lugar paradójico. De un lado está invisibilizada en el universo del conflicto armado: en la reflexión académica no llega ni al estatuto de capítulo independiente, en los medios de comunicación aparece sólo cuando la guerra la asalta y en las políticas públicas hacia la paz apenas si se menciona –como bien lo vuelve a ratificar la actual negociación en la Habana–. Del otro lado la ciudad es objeto de atención al interior de ella misma, así como lo ponen de manifiesto los sentimientos de inseguridad capturados en los sondeos de opinión realizados cada tanto entre sus pobladores. Tal atención, no obstante, no conecta la ciudad sino consigo misma, enlazada por excepción con aquello que tome cuerpo fuera de los confines del perímetro urbano. Cada ciudad construye su propia visión acudiendo a la noción, no de conflicto o guerra, sino de seguridad.

En este complejo entrecruzamiento de miradas se mueve la ciudad: desconocida y marginada frente al conflicto armado, atrapada y circunscrita a la seguridad. Es preciso resituar la ciudad. Primero poniendo en evidencia su singularidad, mostrando que ella produce violencias cuyas gramáticas no se reducen al conflicto armado pero que con todo aportan más de la mitad del homicidio nacional. Segundo, una vez captada su singularidad, conectando la ciudad a las dinámicas del conflicto, tanto de la nación en su conjunto como de la región donde cada ciudad se inscribe.

Resituar la ciudad, así las cosas, interpela las formas de representación del conflicto. De un lado el olvido de lo urbano se produce sobre la reducción del conflicto violento al conflicto armado, todo lo cual viene cargado de consecuencias. El foco centrado en el actor armado, en su palabra y sus estrategias, suprime otros escenarios y otras violencias (como la ciudad y su experiencia), confinando además la sociedad al papel de mero actor pasivo al que no le cabe responsabilidad alguna sobre el curso de la violencia. Del otro lado, en la orilla opuesta, el confinamiento en la seguridad emborrona la ciudad en general –la Ciudad con mayúscula–, haciendo imposible su conexión con algo más allá de la situación doméstica de cada urbe. La noción de seguridad, extendida por el mundo entero, ha de sufrir acomodos una vez busca interpretar un país con un conflicto armado de más de 60 años de permanencia.

Resituar la ciudad conecta también con las prácticas colectivas adoptadas a fin de conjurar el conflicto. Aquí la ciudad vuelve a ser objeto de doble mella. En los muchos esfuerzos puestos en marcha desde la década de los 80 la paz parece no conferirle papel alguno a la ciudad –de nuevo,

3 Se entiende por ciudad la aglomeración de más de 100 mil habitantes. El criterio adoptado se explica más adelante.

como lo ratifica el contenido del diálogo en la Habana–; a cambio, otra vez, los emprendimientos urbanos frente a las violencias se circunscriben a la iniciativa particular del gobierno local. Conflicto, paz y ciudad parecen escindirse, plantando un inquietante interrogante al proceso de paz en Colombia: ¿la paz estable y duradera no pasa por la ciudad, esa gran productora de violencias?

Con el interrogante en mente emprendemos el trabajo de ponerle sustento a más de una afirmación lanzada en esta introducción. Antes de nada es indispensable visualizar la creciente urbanización de la que es objeto hoy el país, haciendo notar la magnitud de sus violencias. ¿Cómo no pensar la paz desde y hacia la ciudad en una nación cada vez más urbana? (apartado I). Luego se enfrentan los vínculos entre conflicto armado y ciudad. ¿Dónde se ubica la ciudad en la conciencia pública sobre el conflicto y qué tanto avanza la guerra sobre la ciudad? (apartado II). Después se aborda la contextualización de lo urbano. ¿Qué nexos mantiene la ciudad con las dinámicas de los conflictos en lo nacional y lo regional? (apartado III). Una vez practicado el recorrido de doble vía entre conflicto y ciudad se ingresa en una primera imagen de la singularidad del conflicto violento urbano. ¿Cómo se conecta la singularidad urbana con el conflicto que está más allá de la ciudad? (apartado IV). Por último en el epílogo, frente a los múltiples estrechamientos de los que es objeto la ciudad, se sugiere la noción de conflicto violento como especie de antídoto contra la reducción y el olvido. ¿A qué nuevas dimensiones de la paz abre la tarea de resituar la ciudad –como veremos la equidad y la convivencia–, una vez cerrado el conflicto armado?

I. Urbanización y violencias

El mundo está marcado por una poderosa tendencia a la urbanización. Si en el año de 1950 habitaba la ciudad el 29% de la población mundial, para el año de 1980 había ascendido al 39% y para el 2010 llegó a ser más de la mitad, el 52% (United Nations Organization, 2011a). El proceso varía de manera considerable de una región a otra. En el 2010 en Africa y Asia estaba urbanizado el 39 y el 44% de su población, al tanto que en Europa y América del Norte lo hacían el 73 y el 82%[4]. América Latina hace parte del impulso urbanizador con ritmos pronunciados. En 1950 en América del Sur hacía parte de la ciudad el 43% de sus habitantes, subió al 67% en 1980 y alcanzó el 83% en 2010. Colombia no es la excepción. En 1950 tenía el 33% de sus pobladores en las urbes, en 1980 llegó al 62% y en 2010 ascendió al 75%. Como se afirmó las tres cuartas partes de la población colombiana habitan la ciudad.

4 En 1950 Africa y Asia estaban urbanizadas en un 14 y 18%; Europa y América del Norte en un 51 y 64%.

Pero, ¿cómo entender la ciudad? Según el *World urbanization prospects*, sobre una muestra de 231 países se identifican cuatro atributos en su definición: administrativos, económicos, población/densidad y características urbanas, combinados de maneras variables de un país a otro (United Nations Organization, 2011b)[5]. En el caso colombiano la definición se adopta sobre la separación de cabecera y resto acuñada por el DANE, una división hecha de la combinación de criterios administrativos y características urbanas (DANE, 2005)[6]. Emplearemos otro criterio[7]. Se entiende "ciudad" como toda aglomeración de 100 mil o más habitantes, una frontera enérgica que disipa cualquier duda respecto a su definición: el número de 100 mil personas incorpora, por la fuerza de la aglomeración, las características que definen lo urbano, esto es grandes conglomerados de personas que habitan territorios contiguos dotados de niveles de densidad residencial.

Para propósitos comparativos se establecerá una diferencia entre ciudad y poblado, separados sobre el mojón de los 100 mil habitantes. Cada uno de los dos, a su vez, se divide en cuatro categorías. La ciudad contiene Megalópolis con más de 4 millones de pobladores; Ciudades Grandes con habitantes entre 1 y 4 millones; Ciudades Intermedias entre 300 mil y un millón; y por último Ciudades Pequeñas de 100 a 300 mil personas. Dicho en breve en el siguiente Cuadro.

Cuadro No. 1
Categorización de ciudades y poblados

CIUDAD	
Megalópolis	Más de 4 millones
Grande	1 millón – 4 millones
Intermedia	300 mil – 1 millón
Pequeña	100 mil – 300 mil
POBLADO	
Grande	50 mil – 100 mil
Intermedio	20 mil – 50 mil
Pequeño	10 mil – 20 mil
Menor	Menos de 10 mil

Validos de la clasificación, la comparación entre México y Colombia confirma la mencionada tendencia a la urbanización (Cuadro No. 2).

5 El informe adopta los criterios de cada país para diferenciar lo rural de lo urbano.

6 "Cabecera" es un área definida por un perímetro urbano establecido por el Concejo Municipal, donde se ubica la sede administrativa del municipio. "Resto" supone la inexistencia de características como calles e instalaciones de servicios públicos al modo urbano.

7 Un trabajo reciente coloca en duda la división cabecera y resto del DANE, mostrando que lo rural tiene una mayor extensión de la reconocida oficialmente (PNUD, 2011).

Ya en 1990 la ciudad congrega más de la mitad de los habitantes, 52% en Colombia y 54% en México; dos décadas después esos porcentajes han crecido a 59% en Colombia y 64% en México. El poblado, por el contrario, disminuye en los dos países. La urbanización se produce hacia arriba de la pirámide urbana, esto es sobre las ciudades[8].

Cuadro No. 2
Porcentaje de población en ciudades y poblados
Colombia y México 1990-2010

		% POBLACIÓN	
PAÍS		**1990**	**2010**
COLOMBIA	Ciudad	52	59
	Poblado	48	41
MÉXICO	Ciudad	54	64
	Poblado	46	36

Fuente: Colombia DANE. México Secretaría de Salud. Cálculos propios

Si se trata de países sometidos a intensos procesos de urbanización, ¿qué sucede con la violencia? Ninguna afirmación unificada puede ser lanzada al respecto, el homicidio de la ciudad se comporta de manera diferente en cada país[9]. En Colombia sigue una curva variable, en unos años la ciudad está por encima del poblado y en otros por debajo (Gráfica No. 1). La información del homicidio entre los años de 1980 y 2010 arma una serie histórica de duración media donde es factible visualizar su comportamiento a lo largo de tres décadas. En contravía de lo esperado –como se encarga de repisar la mirada centrada en el conflicto armado–, las ciudades mantienen tasas de homicidio equiparables (y en largos tramos más altas) respecto a los poblados. Durante buena parte de los años 80 la ciudad y el poblado son casi iguales, hasta cuando en el año de 1988 la ciudad comienza un ascenso vertiginoso que la coloca por encima hasta el año 2000. Durante los siguientes ocho años el poblado permanece un poco arriba, momento en que la ciudad repunta colocándose de nuevo en valores más altos.

[8] El proceso tiene énfasis distintos en cada país. En México crecieron en particular las ciudades grandes con una razón de crecimiento poblacional equivalente al 129%; en Colombia las ciudades intermedias con una razón de 72%.

[9] En Brasil la ciudad está siempre por encima de los poblados mientras en México está siempre por debajo.

Grafico No. 1
Tasa de homicidio de ciudades y poblados. Colombia 1980-2010

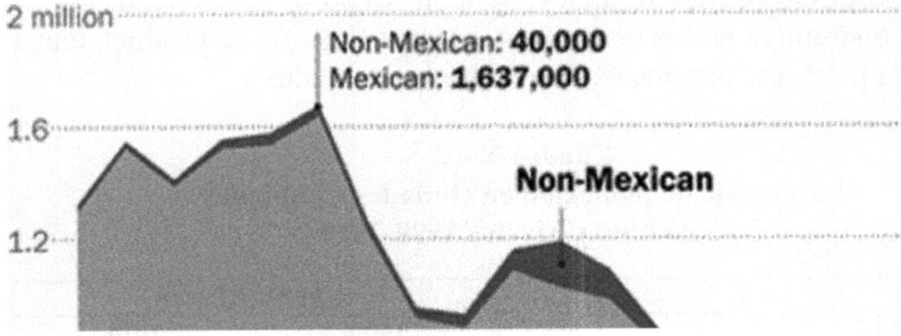

Fuente: DANE (1980-2010). Cálculos propios.

Vista la situación desde el número de homicidios se ratifica la elevada participación de la urbe. En cada una de las tres décadas permaneció igual o por encima del poblado (Gráfica No. 2): en los años 80 puso el 50% del total de homicidios, en los 90 el 61% y en la primera década del 2000 el 54%. En abierta oposición con una vehemente literatura empeñada en desconocer el homicidio de la calle los datos muestran[10], con suficiencia, el peso de la ciudad sobre el período crítico de la violencia reciente de Colombia. La ciudad pone más de la mitad de los homicidios de los últimos 31 años, el 56% como se afirmó al comienzo de la Introducción.

Grafico No. 2
Homicidios en ciudades y poblados por décadas (%)
Colombia 1980-2010

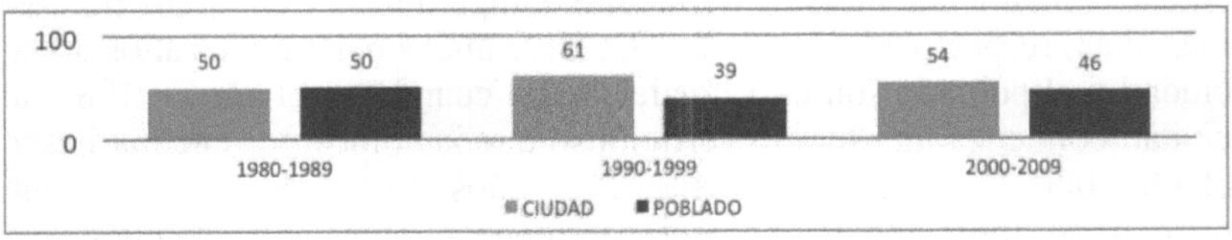

Fuente: DANE (1980-2010). Cálculos propios.

Tal participación de la ciudad no proviene del mayor número de pobladores que alberga, como si bastara con afirmar que la violencia se produce ante todo en el contexto urbano. La relación entre participación en el homicidio y participación en la población lo muestra (Gráfico No. 3). A lo largo de las dos décadas recientes las ciudades de Colombia tienen una participación en la población siempre por debajo de la

[10] El desconocimiento del papel de la ciudad en el homicidio en Echandía (1999) y Rubio (1999).

participación en el homicidio. Dicho en breve, ponen más homicidios que población. No es el caso de México, donde la contribución al homicidio se mantiene por debajo de la contribución a la población otra vez hasta 2007, cuando la contribución al homicidio se pone por encima. Ambos países experimentan una urbanización de igual magnitud, mas la relación homicidio-población es diferente.

Grafico No. 3
Participación de la ciudad en el homicidio y la población. Colombia 1990-2010

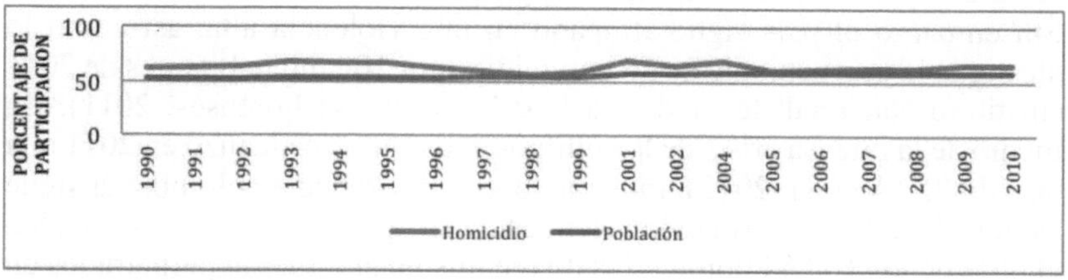

Fuente: DANE. Cálculos propios.

El panorama de la violencia urbana colombiana se completa echando mano de un indicador adicional, las lesiones personales. Haciendo uso de informaciones reportadas en *Forensis* por el Instituto Nacional de Medicina Legal y Ciencias Forenses, esta vez con una serie más corta, de nuevo se revela el mayor protagonismo de la ciudad frente al poblado (Gráfico No. 4): entre 2005 y 2010 la tasa de lesiones personales de la ciudad está siempre por encima. Si la curva de la ciudad no baja en ningún año de 30, la de los poblados no pasa en ningún momento de 23.

Grafico No. 4
Tasa de lesiones personales en ciudades y poblados. Colombia 2005-2010

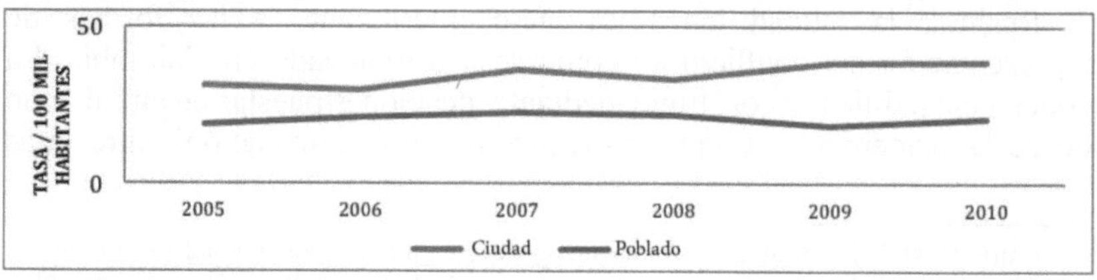

Fuente: Instituto Nacional de Medicina Legal y Ciencias Forenses. Cálculos propios.

En Colombia la ciudad desempeña un papel crucial en el acontecimiento violento –lo revelan el homicidio y las lesiones personales–, afirmación que no cumple del mismo modo en México. Las diferencias entre las naciones ponen en alerta frente al intento de atribuir la violencia urbana a un efecto sin más de la aglomeración. Bogotá lo confirma. Siendo una ciudad en franca expansión, sus tasas de homicidio descienden de manera sostenida a lo largo de los últimos veinte años[11]. Es preciso buscar explicaciones alternativas, unas que den cuenta tanto de las dinámicas propias de lo urbano como de su inscripción en el conflicto nacional.

II. Ciudad y conflicto armado

La intensidad de la violencia en Colombia ha disminuido una vez se la compara con años anteriores, en particular con la década del 90. Sin embargo el país sigue atrapado en una violencia alta, así como lo atestigua la tasa nacional de 36 homicidios por 100 mil habitantes de 2011 (Instituto Nacional de Medicina Legal y Ciencias Forenses, 2011). En medio de la intensa crisis de los últimos años México alcanzó en 2011 una tasa de 25 (Azaola, 2012), mostrando que la violencia colombiana sigue siendo elevada. Confirmando la tendencia en operación desde los años 80, el 65% de los 16.554 homicidios de ese mismo año fueron perpetrados en aglomeraciones de más de 100 mil habitantes[12]. La participación urbana subió, dando cuenta de la manera como el relativo declive del conflicto armado deja en la ciudad una violencia de elevadas proporciones[13]. Ante la arremetida urbanizadora y el ingente homicidio urbano –dos tendencias presentes desde hace 30 años–, la paz enfrenta otro complejo dilema al que tendrá que dar alguna respuesta: el procesamiento de la violencia en la ciudad es una pieza clave de la paz estable y duradera. En este contexto, el primer paso supone hacer claridad sobre los nexos entre ciudad y conflicto armado dando cuenta de dos fenómenos: de un lado la conciencia pública sobre el conflicto, del otro el desarrollo de la guerra misma. Frente a la una y el otro, ¿dónde está la ciudad?

1. La conciencia pública

Resituar la ciudad lanza un recio interrogante a las formas de representación del conflicto tal como se han moldeado en Colombia. La conciencia pública se construye mediante ideaciones puestas en circulación desde la academia, el estado y los medios de comunicación, entre otras

[11] Entre 1990-2010 Bogotá tuvo una razón de crecimiento poblacional del orden del 49%, la más alta entre las ciudades capitales de Latinoamérica. Una mirada sobre Bogotá y su descenso comparativo en la tasa de homicidio en Perea y Rincón (2014).

[12] Medicina Legal reporta información del 82% de los municipios del país con un cubrimiento del 99% de la población.

[13] Decimos relativo declive en la medida en que, para el año 2014, las FARC han perdido considerable presencia y los paramilitares pasaron por una negociación que desmontó parte de su aparato armado.

mediaciones. Como veremos en los tres ámbitos la ciudad tiene un lugar precario, por decir lo menos: en cada uno la lógica de comprensión se construye desde el actor armado de la guerra, produciendo la reducción del conflicto violento al conflicto armado y, como consecuencia, achatando las posibles maneras de comprender el conflicto.

En primer término en la academia. Sus estudios visibilizan realidades y generan categorías de representación de la sociedad y sus procesos. Colombia ha acumulado un rico y variado saber sobre la violencia, así como lo ponen de manifiesto diversos balances bibliográficos[14]. Entre los vacíos que pueden endilgarse a esa producción académica figura con prominencia la ciudad, en ningún balance llega a constituir un tema con identidad reconocida[15].

Dos grandes tipos de estudios se han propuesto una visión comprehensiva del conflicto, unos comprometidos en una visión panorámica, otros empeñados en una interpretación global[16]. Los primeros han hecho de los mapas y la estadística su herramienta preferida, al tanto que los segundos se empeñan ante todo en la interpretación conceptual[17]. En unos y otros, no obstante, la ciudad no constituye nada parecido a un capítulo independiente digno de mención autónoma. En algunos momentos se le menciona –es imposible no hacerlo[18]–, pero en ningún caso hasta el grado de alcanzar una elaboración independiente.

Un libro tras una mirada totalizante de la violencia, *Violencia política en Colombia* (González, Bolívar y Vázquez, 2006), revela bien el silencio en el que permanece sumida la violencia urbana. El horizonte de larga duración en que se inscribe la reflexión obliga a los autores a pasar por una compleja multiplicidad de factores determinantes del prolongado conflicto nacional. El nunca resuelto conflicto agrario, los perversos modos de conexión política entre las regiones y la nación, el partidismo recalcitrante y el narcotráfico, las decisiones conscientes de los actores y la búsqueda de rentas, entre otros, aparecen como resortes desde los cuales la violencia gana presencia histórica en la tramitación del conflicto.

14 Entre los mejores balances González, Bolívar y Vásquez (2006), una analítica que incluye la violencia de mediados del siglo XX; otro excelente desde los años 90 en Peñaranda (2007).

15 La ciudad tampoco aparece en recopilaciones que intentan dar cuenta de las dimensiones del conflicto. Leal y Zamosc (1990), Leal (1995), Leal (1996).

16 Entre los estudios panorámicos destacan Daza (1995), Cubides, Olaya y Ortiz (1998) y Echandía (1989). Entre los interpretativos descollan González, Bolívar y Vásquez (2006) y Gutiérrez, Wills y Sánchez (2006).

17 La división entre estudios panorámicos e interpretativos es por completo arbitraria; se la usa nada más en términos descriptivos y de exposición.

18 Sea el caso, Cubides, Olaya y Ortiz (1998) la traen a cuento en más de una explicación y Echandía (1989) habla del avance del conflicto armado sobre la ciudad.

En medio de ese considerable intento de síntesis y articulación resulta notable la ausencia del tema urbano, en un doble sentido. En primer término en cuanto no aparece articulado a los factores explicativos de las explosiones violentas. Mientras en la literatura internacional la urbanización es considerada un destacado activador del episodio violento, en Colombia, de manera distinta, la ciudad parece no coadyuvar sino de manera episódica[19]. Como consecuencia, en segundo lugar, la ciudad no constituye un objeto temático en sí mismo. En contraste con las amplias descripciones de los actores y su evolución por las regiones, la ciudad aparece como un territorio ausente cuyo destino resulta indescifrable en medio de ese gran despliegue violento.

Naturalmente el vacío de lo urbano no es un error de apreciación imputable sin más a los autores, ellos hacen parte del pensamiento que hegemoniza la mirada nacional sobre la violencia. Se trata de una corriente de pensamiento que arranca con la investigación sobre los episodios de mediados del siglo XX. Pese a que la gran movilización social de los años 30 y 40 tomó como escenario la ciudad, la irrupción de la violencia a mediados de la década del 40 desplazó el foco de atención hacia los sectores rurales[20]. La mirada se ruraliza marcando distancia con la ciudad. La prolija investigación sobre el período no la contempla, ni dentro de sus determinantes explicativos ni tampoco en sus temáticas de interés[21].

La perspectiva ajena a lo urbano pareciera ser puesta en su sitio con el libro *Colombia: Violencia y Democracia*, un texto producido en la segunda mitad de los años 80 que cumpliera un papel decisivo en la configuración del pensamiento sobre la violencia nacional. En él se afirma que "*mucho más que las del monte, las violencias que nos están matando son las de la calle*" (Comisión de Estudio de la Violencia, 1987, 18). La frase, enunciada en la introducción, no sólo hacía virar la mirada hacia el olvidado contexto urbano sino que además sugería que en él se venía gestando la mayor cantidad de procesos violentos. Sin embargo la afirmación, pese a su contundencia, no halló respaldo en el desarrollo ulterior del libro. Por un lado, pese a que el texto se empeña en dar cuenta de la multiplicidad de las violencias, algo así como la violencia urbana no constituye un capítulo independiente; por el otro la parte final, encaminada a formular las recomendaciones de política pública, mencionó el problema pero no a la manera de un espacio que ameritara una intervención en función de su especificidad.

19 Como ejemplo del papel de la ciudad en la violencia y el crimen mirar Centro Internacional para la Prevención de la Criminalidad (2010), un informe mundial.

20 La violencia se volcó al campo una vez neutralizadas las formas organizativas presentes en la ciudad: los sindicatos y la movilización gaitanista. Pécaut (1987).

21 Un balance de la Violencia de mediados del siglo XX en Sánchez (2007), donde resulta igual de visible la ausencia del escenario urbano.

Algunos años más tarde hace irrupción otra interpretación, de todo a todo en oposición con las conclusiones de la Comisión de 1987[22]. La violencia no tiene que ver ni con la inequidad ni con un estado precario y ausente, se decía; el fermento de la violencia ha de buscarse más bien en la poderosa presencia de actores criminales que actúan racionalmente en la consecución de sus intereses. La discusión sobre lo urbano debiera entrar en la analítica sobre la extensión de la violencia en el tiempo y la geografía[23]. En lo atinente a esta última se afirmó que, puesto que las localidades más violentas vienen a ser pequeños municipios rurales donde apenas si clasifica la atormentada Medellín, la afirmación de "*las violencias que nos están matando son las de la calle*" se acusó de carente de fundamento científico y, por ende, producto de la posición ideológica de quienes la formularon. Con dicho aserto lo urbano entra en un callejón sin salida. Qué se mira y cómo se mira, es el punto álgido de discusión una vez se toca el tema de la configuración de la opinión pública: mientras se hacían estas afirmaciones la violencia de la ciudad, como se vio, desbordaba con creces la de los poblados[24]. La delimitación de la práctica violenta a unos actores criminales que actúan en la persecución de sus intereses materiales –sin atribución de racionalidad social alguna–, se llevó hasta sus últimas consecuencias durante los ocho años del gobierno Uribe. La ciudad desaparece del debate sobre el conflicto armado.

Naturalmente la ciudad tiene estudios, algunos de los cuales abrieron un camino que la investigación posterior tendrá que seguir recorriendo[25]. Dos grandes momentos históricos han creado, cada uno en su contexto, un respectivo pico de producción académica sobre lo urbano. El primero fue la crisis de Medellín, cuando el cartel de Pablo Escobar y su empleo sistemático de sicarios llevaron la violencia a niveles insospechados[26]. Es la ciudad donde mayor desarrollo ha alcanzado la reflexión sobre su violencia interna, en buena medida como resultado de una conciencia académica forzada a pensar una profunda tragedia. Los temas son

22 El texto más ilustrativo de la argumentación de los llamados economistas en Montenegro y Posada (1995).

23 La violencia no cruza la totalidad de la historia (el tiempo) y tampoco se expande por igual a lo largo del territorio, dicen los economistas.

24 El libro de la Comisión salió en 1987, un año antes del escalamiento de la violencia urbana. Los economistas lanzan su crítica a mediados de los años 90, momento en que la urbe tiene de lejos una tasa de homicidio superior. Volver al Gráfico No. 1.

25 La relación ciudad y conflicto arranca con el trabajo pionero de Alvaro Camacho y Alvaro Guzmán (1990), primero que convierte la violencia urbana en objeto de reflexión a partir de la situación de Cali. De varios modos esta investigación influyó los planteamientos de la Comisión de 1987.

26 Ana María Jaramillo (2011) hace un excelente balance de la literatura sobre Medellín.

indicativos –narcotráfico, sicariato, combos de barrio[27]–, todo en el contexto de una violencia que ha dado origen a sugerentes desarrollos temáticos[28].

El segundo momento histórico se constituye alrededor de la invasión paramilitar de la ciudad. Como lo anunciara Carlos Castaño en 1999, sus fuerzas emprendieron de ese momento en adelante una toma de las ciudades que alcanzó importantes grados de penetración en varias de ellas. Se posesionaron en zonas populares controlando la circulación de las rentas ilegales, extorsionando los negocios locales y administrando la violencia[29]; impusieron su dominio sobre negocios estratégicos como Corabastos y los Sanandrecitos en Bogotá, al tiempo que hegemonizaron empresas como las apuestas y los juegos de azar; en algunos casos tomaron por la fuerza el control de prominentes negocios legales, como fue el caso de los gremios lecheros en la Costa Caribe; y en algunas otras ocasiones detentaron el control de las arcas de los estados locales[30]. Hay una producción sobre lo urbano; empero, pese a sus notables avances, ella no alcanza a resolver el olvido de la ciudad en la reflexión sobre la violencia.

La academia no es la única mediación de la conciencia pública donde la ciudad se invisibiliza, otro tanto acontece con los medios de comunicación[31]. Durante los últimos 25 años existen dos momentos en que lo urbano cobra vida en el interés de las informaciones de los medios[32]. El primero se produce cuando un actor armado ejecuta incursiones de cierta espectacularidad sobre la ciudad o sus alrededores; el segundo, más escaso, cuando el conflicto armado parece encontrar alguna forma de "solución"[33].

---

27 Entre los más destacados Salazar (1990), Salazar y Jaramillo (1992), Jaramillo, Ceballos y Villa (1998), Bedoya y Jaramillo (1991).

28 Elsa Blair (2004).

29 Duncan (2005) hace un sugerente texto en que mira la ciudad como una totalidad. Una versión de este artículo aparece en su libro *Los señores de la guerra*.

30 Textos sobre paramilitarismo y ciudad en Pérez (2007), Espinal, Giraldo y Sierra (2007), Avila y Núñez (2009), Restrepo (2009), Grupo investigación Casa de Paz (2009).

31 Lo que sigue es apenas un esbozo. Las limitaciones de espacio impiden una elaborada reflexión sobre la ciudad en los medios.

32 La afirmación se construye sobre la revisión de los artículos producidos por la revista Semana a partir del año 1985. La base fue construida por Sayra Rodríguez, Alejandro Pajón y Paola Ocampo para la celebración de los 25 años del Instituto de Estudios Políticos y Relaciones Internacionales (IEPRI).

33 Nos ocupamos tan sólo del primer momento. Ejemplo del segundo momento es la crispada preocupación sobre la ciudad al comienzo del gobierno Pastrana cuando la guerra parece hallar solución en las negociaciones con las FARC, el ELN y hasta los paramilitares.

En 1985 apareció un artículo bajo el título de la "La guerra llega a los barrios", seguido de un sugestivo subtitulado que rezaba "más que en el campo, es en las ciudades donde se juega la suerte de la paz" (Semana, junio 17 de 1985). La afirmación se armaba sobre la ocurrencia de algunos eventos que por ese entonces enervaban los ánimos[34]. Los hechos y la preocupación que generan tienen legitimidad, no cabe duda; el punto es, no obstante, la lógica mediante la que se construye el texto: la ciudad hace su incursión en los medios conectada a esos momentos en que "la guerra llega a los barrios". Así lo hará durante los siguientes 25 años. Los campamentos de paz visibilizarán a Cali y Bogotá a mediados de los años 80; Pablo Escobar con sus particulares y en extremo sangrientos modos de actuación destacarán a Medellín a finales de la misma década; el cartel de los Rodríguez Orejuela producirá la presencia mediática de Cali en los primeros años de los 90; la presencia de las FARC en inmediaciones de la capital pondrá en primer plano a Bogotá entre 1996 y 2001; la incursión paramilitar evidenciará algunas ciudades a comienzos del 2000. Cada tanto un notable acontecimiento –nunca faltan– sube a la palestra una ciudad, como el secuestro masivo de feligreses en la iglesia de la María y pocos días después el secuestro masivo de diputados en Cali ("Por qué Cali". Semana, octubre 23 de 2000).

Ciertamente la ciudad accede a las informaciones de los medios de comunicación. Con todo, ella habla a través de su conexión con el conflicto armado, despojada de su singularidad. En realidad los medios posan su atención sobre el conflicto armado, la ciudad es la circunstancia. Naturalmente no se trata de una circunstancia despreciable, el proyecto estratégico de la insurrección se corona con la toma de la ciudad. De allí que, por lo general, la noticia urbana se cargue de aseveraciones grandilocuentes como aquella recién citada de "más que en el campo, es en las ciudades donde se juega la suerte de la paz". Como acontece con el universo de la academia, también en el de los medios de comunicación el conflicto armado engloba y subsume el conflicto violento opacando sus otras facetas. La reducción del conflicto violento al conflicto armado, no obstante, se produce de modo distinto en cada mediación. En la academia como olvido, en los medios en cuanto sólo la guerra parece hablar de la ciudad[35].

---

[34] El Ricardo Franco incursionó sobre la escuela de carabineros de Suba en Bogotá, el ELN practicó un asalto sobre la cárcel Modelo, el M-19 organizó campamentos en barrios populares.

[35] Naturalmente aparecen de cuando en vez noticias de aspectos "propios" de la ciudad. Por ejemplo sobre escuadrones de la muerte "Con sus propias manos". Semana, mayo 19 de 1986; en torno a los skinheads "¿Quiénes son estos calvos?". Semana, junio 28 de 1993; o sobre el crimen organizado "Se alquilan para matar". Semana, noviembre 3 de 2002.

La precariedad de lo urbano, en tercer y último término, se revela aún más todavía en el terreno de las políticas públicas respecto a la violencia y la paz. En ninguno de los esquemas globales hasta hoy adoptados el asunto urbano merece un tratamiento particular. La tesis de la conexión entre pobreza, ausencia del estado y violencia, que rigiera las administraciones de Belisario Betancur y Virgilio Barco entre 1982 y 1990, presuponía la inversión en zonas rurales apartadas con el fin de cerrar la brecha de exclusión y aislamiento que sirviera de caldo de cultivo a la violencia[36]. La *Política Integral contra la Violencia*, enunciado con el que César Gaviria (1990-1994) tituló su política pública una vez sancionada la Constitución del 91 y una vez fracasadas las conversaciones de Caracas y Tlaxcala, tampoco contempló medida alguna en la dirección de visibilizar la ciudad; el calificativo de *integral* presidiendo una política donde se soslaya lo urbano, da buena cuenta del precario lugar que ocupa el homicidio de la calle en las preocupaciones colectivas sobre la violencia. Lo confirmarán luego Pastrana y Uribe entre los años de 1998 y 2010[37]. El primero al centrar su política en la zona de despeje y la negociación con los alzados en armas, el segundo al focalizar el esfuerzo estatal en la persecución y aniquilamiento de la FARC en sus zonas de apoyo y abastecimiento en el campo[38].

En estos términos, toda vez que se aborda el conflicto armado la ciudad se desdibuja. Sucede en la academia, pero también acontece en los medios de comunicación y las políticas estatales. Es un modo establecido de la conciencia pública: la guerra es la que cuenta. Por supuesto el privilegio del conflicto armado tiene razones de fuerza histórica y política, la ascendencia de la guerra sobre el curso de la vida colectiva es impresionante. Los actores armados controlan un sinnúmero de procesos de la vida de las gentes que habitan los lugares donde se implantan, llegando al extremo de administrar justicia sobre asuntos granados de la vida de todos los días; detentan poder sobre los estados locales y los fiscos municipales, pasando por la imposición de extorsiones a toda obra de beneficio público; poseen empresas económicas de grandes rendimientos como el narcotráfico, mientras someten multitud de empresas grandes y pequeñas al pago de cuotas forzadas. Entretanto, mientras la guerra sigue su curso, Colombia ha vivido uno de los desastres humanitarios más críticos del planeta, atiborrado de desplazados, desaparecidos, torturados, ejecutados, asesinados … La esfera pública de las últimas décadas no ha

36 Una mirada sobre la década en Leal y Zamosc (1990).

37 Samper, ocupado como estuvo en el asunto de los dineros del narcotráfico en su campaña, no tuvo lugar para enunciar alguna política global de paz.

38 La única referencia explícita en Uribe será la guerra de Medellín durante la segunda mitad de 2002, recién subido a la presidencia.

podido evitar ser construida en conexión con la existencia del conflicto armado, desde los grandes emprendimientos institucionales como la constitución de 1991 hasta las masivas inversiones regionales al estilo de los planes de rehabilitación. Las ocho elecciones presidenciales de 1982 a hoy han discurrido todas bajo el signo de la guerra y la paz, del conflicto armado y el intento de exorcizarlo.

Frente al peso abrumador de la guerra la paz pasa, en primera instancia, por la desactivación del conflicto armado que sobreviene con la desmovilización de los combatientes irregulares. No hay la menor discusión al respecto. No obstante, la importancia histórica de la guerra no suprime sus complejas implicaciones sobre las formas de interpretación e intervención del conflicto. El propósito de erradicar la práctica de la violencia –de reducirla a términos internacionales razonables[39]–, halla en la reintegración civil de los alzados en armas uno de sus notables logros; empero, el diálogo con las guerrillas está lejos de resolver una enorme cantidad de realidades del conflicto violento, buena parte de ellas incrustadas en la calle atendiendo a las gramáticas del asfalto y la congestión. La desactivación de la violencia urbana es un nodo crucial del establecimiento de las condiciones que garanticen una paz estable y duradera.

2. Conflicto armado y ciudad

La invisibilidad de lo urbano en la conciencia pública del conflicto se vincula, de buen modo, al hecho de que la guerra ingresa sólo de manera parcial a la ciudad. En ningún caso se sugiere que la guerra no llegue a la calle. El escenario urbano es un engranaje primordial de la confrontación, no sólo porque la ciudad desempeña una función estratégica en la inteligencia militar, el avituallamiento de los ejércitos y la ampliación de la base social de la guerra, sino porque la toma de la ciudad encarna el triunfo de la lucha insurgente. Quien tiene la ciudad tiene el poder. De allí que el plan estratégico de las FARC pasó por el intento de cercar la capital Bogotá. El ingreso recortado de la guerra a la ciudad significa entonces que, salvo contadas excepciones[40], los actores de la guerra no consolidan allí un dominio ni de cerca parecido al que logran implantar en zonas rurales, donde imponen leyes de estricto cumplimiento, administran justicia, cobran "impuestos" y en general imponen su voluntad. Así pues,

[39] La tasa mundial de homicidio es de 8,8. Organización Panamericana de la Salud-Organización Mundial de la Salud (2003).

[40] Son tres las excepciones. Primero Medellín, de la que se hablará en el último apartado; segundo Barrancabermeja, escenario de confrontación entre guerrillas y paramilitares; y tercero un momento de la expansión paramilitar a comienzos de la década del 2000, cuando hicieron presencia en varias ciudades.

¿hasta dónde llega la guerra a la ciudad? Abordamos el interrogante acudiendo a cinco indicadores del conflicto armado: el desplazamiento, el secuestro guerrillero, las acciones bélicas, la violación del derecho internacional humanitario y la violación de derechos humanos.

Primero el desplazamiento forzado, un prominente indicador del desastre humanitario en cuanto él supone el abandono del lugar de origen dejando la vida y los bienes al arbitrio de los victimarios. La información de Acción Social arma el cuadro de las personas expulsadas entre los años de 1998 y 2011 (Gráfico No. 5)[41]. Con entera claridad desde el poblado se genera el gran monto del desplazamiento. En el año de 1998 produjo el 94% y en 2011, tras un sostenido descenso, sigue produciendo el 61% de las expulsiones. La ciudad se comporta de manera inversa. En 1998 fue asiento de un pequeño 6% de las expulsiones pero tras una curva de ascenso llegó al 39%. Si bien el desplazamiento desde la ciudad cobra proporciones alarmantes –llega a ser casi las dos quintas partes en 2011–, el mayor volumen ha sido y sigue siendo producido desde el poblado.

Grafico No. 5
Participación de ciudades y poblados en el desplazamiento. Colombia 1998-2011

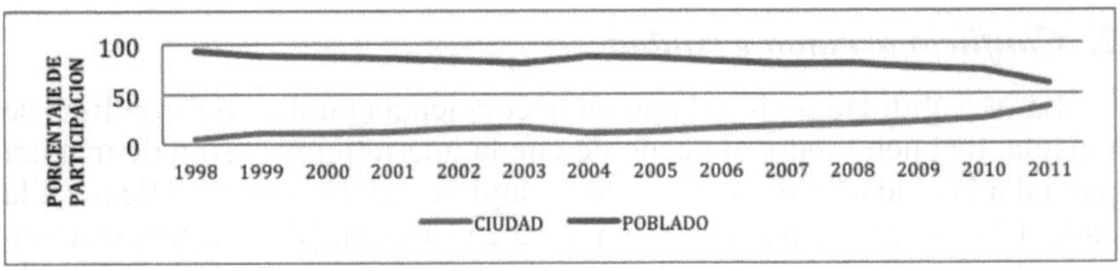

Fuente: Acción Social. Cálculos nuestros.

El seguimiento de los restantes indicadores del conflicto armado acude a la revista *Noche y Niebla* del Centro de Investigación y Educación Popular (CINEP), la más antigua base de datos sobre derechos humanos existente en Colombia[42]. En primer término se mira el 2001, un año de importancia histórica en el curso del conflicto armado en tanto representa un punto culminante de la llamada fase de recrudecimiento del conflicto –desarrollada entre los años de 1996 y 2002– (Restrepo y Aponte, 2009)[43]. Tanto el crecimiento de la guerrilla como el del paramilitarismo tiene en este momento su máximo esplendor; además el proceso de la negociación con las FARC en el Caguán pasa por su tercer año y se prepara el ascenso

41 En línea en <www.accionsocial.gov.co>.

42 También en línea en <www.nocheyniebla.org>.

43 El punto más alto de esta fase de recrudecimiento se alcanza en 2002.

de la administración Uribe, cuyas ejecutorias modificarán el curso del conflicto. Finalmente, y no menos importante para nuestro empeño, el año 2001 coincide con el momento en que la opinión pública habla de la toma urbana por parte de la insurgencia, motivada en el cerco que por ese entonces tienden las FARC sobre Bogotá.

Iniciamos con uno de los delitos más envilecidos, el del secuestro guerrillero. Pese a la posible predicción –cabría esperar que la ciudad con sus recursos fuera epicentro de este tipo de secuestro–, los datos confirman el panorama bosquejado ya en el desplazamiento. En el año 2001, cuando la acción guerrillera alcanza uno de sus picos y el secuestro nacional llega a su segundo punto más alto[44], se comenten en el poblado el 88% de estos delitos (Cuadro No. 3). El 65% toma lugar en los municipios intermedios y pequeños. El cuadro desconcierta, la ciudad es asiento de nada más que 12% de los secuestros ejecutados por uno de los más fuertes actores armados de ese momento.

Cuadro No. 3
Secuestro realizado por la guerrilla. 2001

	CATEGORÍA	RANGO	No.	%
CIUDAD	Megalópolis	Más de 4.000.000	1	0,3
	Grande	1.000.001-4.000.000	1	0,3
	Intermedia	300.001-1.000.000	9	3
	Pequeña	100.001-300.000	31	9
	TOTAL CIUDAD		42	12
POBLADO	Grande	50.001-100.000	34	10
	Intermedio	20.001-50.000	110	32
	Pequeño	10.001-20.000	115	33
	Menor	Menos de 10.000	46	13
	TOTAL POBLADO		305	88
	TOTAL PAÍS		347	100

Fuente: Revista Noche y Niebla. Cinep. 2001

La revista *Noche y Niebla* categoriza la violencia política en tres categorías dependiendo de quién la ejerza y el modo como la ejerza[45]: agentes del Estado o particulares que actúan con aquiescencia del Estado, en cuyo caso se trata de violación de los derechos humanos; grupos insurgentes que se atienen a las leyes y costumbres de la guerra, referida como acciones bélicas; y grupos insurgentes que quiebran las normas que

44 El secuestro extorsivo alcanzó su mayor nivel en el año 2000, cuando se cometieron 3606. En el 2001 se cometieron 2357. Policía Nacional (1990-2010).

45 Para efectos de nuestro interés se excluye el tercer tipo especificado en el marco conceptual (la violencia política social). El marco conceptual en <http://www.nocheyniebla.org/node/18>.

regulan la guerra, tipificadas como violación del derecho internacional humanitario[46]. En primer lugar las acciones bélicas, un excelente indicador del conflicto armado que describe la estrategia militar de las guerrillas, incluyendo 10 tipos de acciones[47]. El desarrollo de la acción insurgente en los poblados es evidente (Cuadro No. 4). En ellos tomó cuerpo el 90% de las acciones bélicas del año 2001, otra vez de modo privilegiado en poblados intermedios y pequeños que agregados sirven de escenario al 66% de las incursiones guerrilleras. Las ciudades sólo presenciaron un 10%, la mayoría en ciudades pequeñas entre 100 y 300 mil pobladores.

Cuadro No. 4
Acciones bélicas en ciudades y poblados (%). Colombia 2001

	CATEGORÍA	RANGO	%
CIUDAD	Megalópolis	Más de 4 millones	0,2
	Grande	1 millón – 4 millones	1
	Intermedia	300 mil – 1 millón	2
	Pequeña	100 mil – 300 mil	6
	TOTAL CIUDAD		10
POBLADO	Grande	50 mil – 100 mil	7
	Intermedio	20 mil – 50 mil	36
	Pequeño	10 mil – 20 mil	30
	Menor	Menos 10 mil	17
	TOTAL POBLADO		90

Fuente: Revista Noche y Niebla-Cinep. Cálculos nuestros

Las violaciones del derecho internacional humanitario, de su lado, incluyen 8 tipos de acciones[48]. Pese a que el listado incorpora amenazas, ejecuciones extrajudiciales, toma de rehenes y otras que bien podrían tomar cuerpo en la ciudad, las violaciones al derecho internacional humanitario tienen lugar, otra vez, en los poblados de menos de 100 mil habitantes (Cuadro No. 5). Los municipios intermedios y pequeños parecen ser el escenario "natural del conflicto": en esta oportunidad agregan el 57% de

46 La división supone que la guerrilla no viola derechos humanos –sólo lo hacen el Estado y sus aliados–, un punto álgido de discusión con la base.

47 Ametrallamiento y bombardeo; ataque a objetivo militar; bloqueo de vías; civiles heridos en acciones bélicas; combate; emboscada; incursión; pillaje; sabotaje; y uso de minas-minado de un campo.

48 Amenaza; ametrallamiento y/o bombardeo indiscriminado; ataque a obras o instalaciones que contienen fuerzas peligrosas, a la estructura vial o bienes indispensables para la sobrevivencia de la población; bloqueo de vías; civiles muertos o heridos en acciones bélicas; colectivo amenazado, desplazado o usado como escudo; ejecución extrajudicial u homicidio sobre persona protegida; perfidia, pillaje, toma de rehenes, tortura, uso de minas y violencia sexual.

las violaciones[49]. La ciudad, entretanto, recoge el 22%, más de la mitad en ciudades pequeñas.

Cuadro No. 5
Violaciones al derecho internacional humanitario (%). Colombia 2001

	CATEGORÍA	RANGO	%
CIUDAD	Megalópolis	Más de 4 millones	2
	Grande	1 millón – 4 millones	2
	Intermedia	300 mil – 1 millón	5
	Pequeña	100 mil – 300 mil	14
	TOTAL CIUDAD		22
POBLADO	Grande	50 mil – 100 mil	7
	Intermedio	20 mil – 50 mil	35
	Pequeño	10 mil – 20 mil	22
	Menor	Menos 10 mil	13
	TOTAL POBLADO		78

Fuente: Revista Noche y Niebla-Cinep. Cálculos nuestros

El panorama es todavía más indicativo en el caso de la violación de derechos humanos. De nuevo, sus acciones incluyen un conjunto de ejecutorias que bien podrían tomar cuerpo en las calles de la ciudad[50]. En efecto el indicador es el más elevado de todos los hasta ahora examinados, en la ciudad tiene lugar el 36% de las violaciones, una vez más con un 21% cometido en ciudades pequeñas (Cuadro No. 6). La anotada prelación de los poblados entre 10 y 50 mil habitantes se mantiene, en esta oportunidad agregando un 47% de los casos[51].

49 Desde estos mismos poblados se produce el 62% de los desplazamientos forzados, confirmando la importancia de los intermedios y pequeños para el conflicto.

50 Amenaza y atentado; colectivo amenazado y desplazado; desaparición y detención arbitraria; ejecución extrajudicial y homicidio sobre persona protegida; tortura; y violencia sexual.

51 La territorialidad del conflicto armado se completa con la presencia de los actores armados: en la ciudad la guerrilla hizo presencia con el 13% de sus acciones, los paramilitares con el 29%.

Cuadro No. 6
Violaciones de derechos humanos (%). Colombia 2001

	CATEGORÍA	RANGO	%
CIUDAD	Megalópolis	Más de 4 millones	4
	Grande	1 millón – 4 millones	3
	Intermedia	300 mil – 1 millón	8
	Pequeña	100 mil – 300 mil	21
	TOTAL CIUDAD		36
POBLADO	Grande	50 mil – 100 mil	8
	Intermedio	20 mil – 50 mil	31
	Pequeño	10 mil – 20 mil	16
	Menor	Menos 10 mil	9
	TOTAL POBLADO		64

Fuente: Revista Noche y Niebla-Cinep. Cálculos nuestros

La situación descrita para el 2001, el de una guerra que llega sólo de manera tangencial a la ciudad, no es el caso de este año en particular. Varios años antes y otros tantos después la distribución geográfica del conflicto armado ha sido y sigue siendo la misma (Cuadro No. 7). Se tomarán como eje de referencia los años de 1997 y 2005, dos momentos diferenciados no sólo por su distancia en el tiempo sino por las condiciones de la guerra en cada uno de ellos. El año de 1997 representa el comienzo de la fase de recrudecimiento de la guerra; el 2005 habla de un instante donde se hacen visibles los éxitos de las operaciones militares del Estado. Pese a las distancias en el tiempo el cuadro dibujado para 2001 se confirma haciendo posible la afirmación de un comportamiento generalizado de la guerra[52]. Ciertamente las acciones bélicas –que en 2001 tuvieron en la ciudad un valor de 10%–, en 1997 son el 12% y en 2005 el 13%, dando cuenta de la inexistencia de variación alguna. La violación del DIH lo mismo –en 2001 fue 22%–, mientras en los otros dos años 17 y 24% respectivamente, igual sin variaciones de consideración. La violación de derechos humanos –en 2001 fue 36%–, en 1997 bajó a 21% y en 2005 subió a 36%; entre los años noventa y el comienzo de la década hubo aumento pero en 2005 se mantiene igual.

[52] Según el Observatorio del Conflicto Violento del IEPRI, en el año 2011 las acciones del conflicto armado disminuyen más todavía en la ciudad. La guerrilla está emplazada en sus antiguas zonas de retaguardia.

Cuadro No. 7
Acciones bélicas, violación DIH y DH (%). Colombia 1997 y 2005

		ACCIONES BÉLICAS		VIOLACIÓN DIH		VIOLACIÓN DH	
	CATEGORÍA	**1997**	**2005**	**1997**	**2005**	**1997**	**2005**
CIUDAD	Megalópolis	0,9	0	0,7	6	2	9
	Grande	0,9	1	2	3	3	6
	Intermedia	3	5	4	9	5	13
	Pequeña	8	7	10	7	12	7
	CIUDAD	12	13	17	24	21	36
POBLADO	Grande	8	9	12	8	13	6
	Intermedio	35	42	33	41	36	37
	Pequeño	30	25	27	17	21	14
	Menor	15	12	11	10	9	7
	POBLADO	88	87	83	76	79	64

Fuente: Revista Noche y Niebla-Cinep. Cálculos nuestros

La ciudad no es el escenario del conflicto armado, sus calles no provocan ni el grueso del desplazamiento ni tampoco la mayor cantidad de secuestros guerrilleros, acciones de guerra y violaciones de derechos fundamentales. Lo urbano encierra una verdadera paradoja: pese a su invisibilidad en la conciencia colectiva en torno a la violencia –en buena medida porque la guerra la alcanza con precariedad repetimos–, aporta más de la mitad de los homicidios nacionales.

III. Seguridad y conexión regional

Si las ciudades enfrentan un fenómeno crítico de violencia pero no son el escenario del conflicto armado, ¿será porque lo urbano se sujeta a una violencia autónoma, reglada por sus lógicas domésticas? Tal la tendencia dominante. La reclusión de la ciudad en la seguridad –tal como se afirmó atrás–, construye, por una vía distinta, otra invisibilidad de la ciudad: la que proviene del ensimismamiento en sí misma. La ciudad ha sido olvidada en la reflexión académica, se señaló; no obstante, se anotó también, dos coyunturas históricas generaron una respectiva oleada de investigación urbana revelando el talante del ánimo imperante. Salvo una excepción todas están centradas en el conflicto de una ciudad en particular[53]. Las comparaciones entre ciudades escasean[54] y menos se

53 La excepción es el trabajo de Duncan (2005 y 2006), quien hace una comparación entre ciudades.

54 El caso paradigmático del ensimismamiento es Medellín. Mucho se ha escrito sobre ella pero las conexiones con municipios como Itaguí y Bello (dos municipios por completo ligados al casco urbano), casi ni son tenidas en cuenta.

avanza en la comparación del conflicto urbano del país con el conflicto de otros países[55].

La ciudad está encerrada, circunscrita a sí misma. Su singularidad, suprimida por el conflicto armado, es llevada al otro extremo, el de la auto contención. Por supuesto la analítica sobre la ciudad en sí misma registra notables avances. No es posible dejar de mencionar las investigaciones de las Cámaras de Comercio de una buena cantidad de capitales, quienes atentas a la demanda de seguridad que entraña el funcionamiento del gremio bajo su atención, habilitan sistemas de vigilancia de los delitos de alto impacto, algunas con elevados niveles de sofisticación[56]. Sin embargo –mostraremos a continuación–, lo urbano tiene más de un nexo con el conflicto ampliado del país. Resituar la ciudad (visibilizarla), generando su inclusión en las políticas públicas sobre la violencia y la paz, supone tanto conectar el homicidio de la calle a la dinámica de la violencia nacional y regional, como reconocer el aporte que entraña su singularidad en la comprensión de la producción del hecho violento. En lo que sigue nos ocupamos de la primera afirmación.

1. El vínculo con lo nacional

Ateniéndonos a la curva del homicidio entre los años de 1980 y 2010, el conflicto violento de Colombia sigue una trayectoria dividida en seis fases (Gráfico No. 6). La primera entre los años de 1980 y 1984, caracterizada por la estabilidad alrededor de una tasa de 30. Durante esos años se da inicio al primer proceso de paz (en 1982) mientras el narcotráfico no irrumpe todavía en la escena pública. Lo hará en mayo de 1984 cuando asesina al ministro de justicia Rodrigo Lara Bonilla en una calle de Bogotá. El efecto se deja sentir de inmediato, la tasa de homicidio se dispara a partir de ese instante siguiendo una línea de ascenso que culmina en el año de 1991, el pico más alto de la historia documentada del país: 86 homicidios por 100 mil habitantes. Es la segunda fase (1985-1991), marcada por la guerra del narcotráfico en el intento de desmontar la extradición, pero también por la carrera de fortalecimiento de las guerrillas y el incipiente desarrollo paramilitar. El proceso constituyente de 1991 marca el punto de inflexión, acompasado por la entrega del cartel de Medellín y la posterior captura de los capos del cartel de Cali. El homicidio desciende hasta 1997 fijando la frontera de la tercera fase (1992-1997), año de un nuevo ascenso prolongado hasta el 2002 como resultado de la expansión territorial y geopolítica de los dos grandes actores armados: la guerrilla de las FARC y la fuerza

---

55 Perea compara ciudades (2007) y también compara con otros países (2008).

56 Un caso ejemplar es el de la Cámara de Bogotá, construida sobre un completo sistema de consulta virtual.

paramilitar de las AUC (1997-2002). La quinta fase, entre 2002 y 2008, se define sobre la caída vertiginosa que produce la estrategia combinada de la seguridad democrática: guerra contra la guerrilla y negociación con los paramilitares. Por último la sexta fase (2008-2010) arranca con el agotamiento de la doble estrategia –la guerrilla se acomoda a la guerra y los paramilitares se remozan en nuevas formas–, abriendo un momento de estabilización con oscilaciones que bordea la tasa de 40.

Grafico No. 6
Tasa de homicidio nacional. Colombia 1980-2010

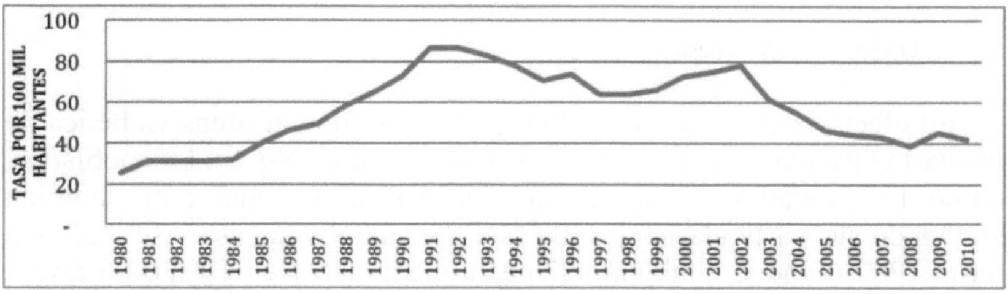

Fuente: DANE. Cálculos nuestros.

Resulta sintomático que la narración recién elaborada considere ante todo el conflicto armado y las respuestas institucionales emprendidas en el intento de conjurarlo, pasando de manera inadvertida la ciudad. No existe un relato de lo urbano con la capacidad de impactar el curso de los acontecimientos nacionales[57]. Por lo pronto impacta el hecho de la profunda compenetración del conflicto nacional con el conflicto urbano, así como se desprende de la comparación de las dos curvas en el Gráfico No. 7. Coinciden casi punto por punto, excepto cuando lo urbano se levanta por encima de lo nacional a lo largo de la década que corre entre 1988 y 1998, año en que la ciudad se pone ligeramente por debajo otra vez durante una década (hasta el 2008). La ciudad sigue paso a paso las tendencias que rigen las seis fases del conflicto nacional, con dos particularidades dignas de mención. En primer lugar la segunda fase no termina en 1991 sino en 1993, aunque el ascenso se detiene igual en 1991; en segundo término la cuarta fase (1997-2002) es menos pronunciada, poniendo en evidencia que el recrudecimiento del conflicto impulsado por el ascenso guerrillero y paramilitar toma cuerpo ante todo en los poblados.

57 Hasta aquí habría que afirmar que el narcotráfico tiene uno de sus asientos en la ciudad, al menos lo fue en la época de los grandes carteles de Medellín y Cali (1984-1993).

Grafico No. 7
Tasa de homicidio nacional y de la ciudad. Colombia 1980-2010

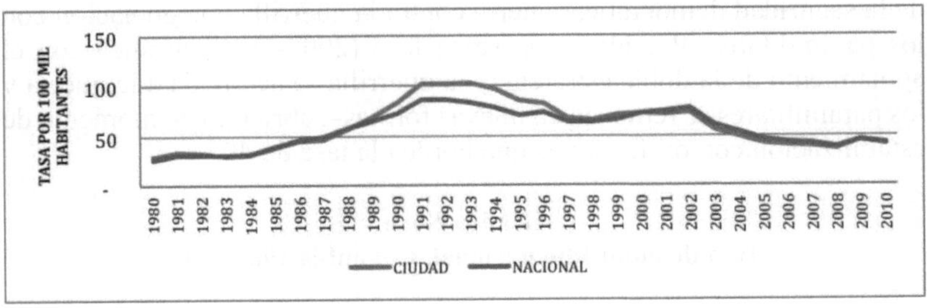

Fuente: DANE. Cálculos nuestros.

En efecto la curva de los poblados muestra que la ofensiva bélica de la cuarta fase realmente comenzó en 1995 (Grafico No. 8). Los poblados, como las ciudades, coinciden con el patrón nacional pero también siguiendo sus particularidades. La fase 2 es menos intensa y se prolonga hasta 1992 (aunque también detiene su crecimiento en 1991). La fase 3 es más corta, la caída se detiene en 1995 como se acaba de señalar, dando inicio a un ascenso de mayor realce que se coloca por encima de la ciudad a partir de 2000. El coeficiente de correlación entre la curva de la ciudad y de los poblados es elevado (0,74), poniendo sobre la mesa una estrecha conexión entre lo que sucede al interior de las ciudades y lo que acontece con el conflicto armado en los poblados[58].

Grafico No. 8
Tasa de homicidio de ciudades y poblados. Colombia 1980-2010

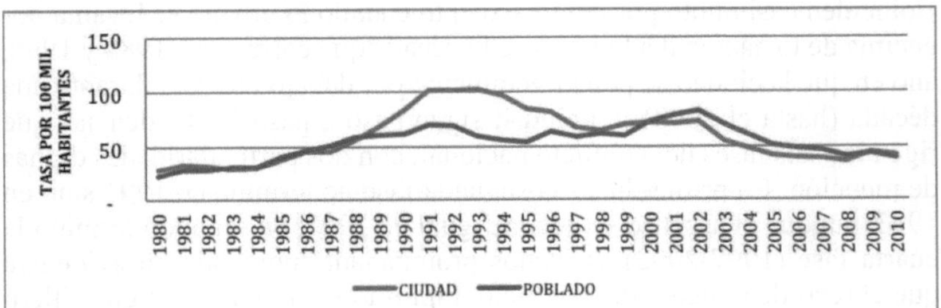

Fuente: DANE. Cálculos nuestros.

[58] Una sugerente trabajo comparando Medellín y el país en Giraldo (2011).

2. *La conexión regional*

La ciudad está conectada a lo nacional, sus grandes tendencias así como sus inflexiones hacia el aumento o la desaceleración coinciden con las fases del conflicto nacional. La ciudad marca el conflicto, éste marca a la ciudad. Otro tanto habrá de aseverarse en torno a la inscripción regional, el conflicto urbano permanece en estrecha relación con sus contextos inmediatos armando definidos patrones regionales. El caso de las ciudades de la subregión Medellín es ilustrativo[59], incluyendo no sólo los municipios de su zona metropolitana (Bello, Envigado e Itaguí), sino también la ciudad de Rionegro donde se encuentra el aeropuerto[60]. El modelo regional salta a la vista, el más fidedigno al modelo nacional de las seis fases (Gráfico No. 9). Naturalmente son varias las especificidades. Desde su intensa magnitud Medellín marca la curva nacional y la curva de las ciudades, su comportamiento se apega a las seis fases excepto por el nuevo escalamiento que arranca en 2008. Las menos coincidentes son Envigado y más todavía Rionegro, las que sin embargo no dejan de atenerse al patrón regional.

Grafico No. 9
Tasa de homicidio zona metropolitana de Medellín y Rionegro 1980-2010

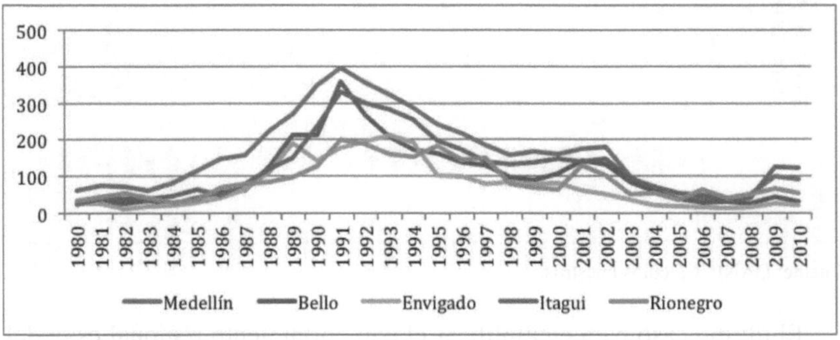

Fuente: DANE. Cálculos nuestros.

La Costa Caribe encarna otro patrón, visible aún entre ciudades sin contigüidad espacial alguna. Las seis fases del homicidio nacional aplican sólo en el caso de Montería, excepto en la sexta fase por cuanto entra en un enervamiento conectado a la presencia violenta de los

59 Entre las 7 ciudades antioqueñas hay dos subregiones, Urabá y Medellín. Hablamos sólo de la segunda.

60 En la definición institucional la zona metropolitana de Medellín incluye varios municipios más; aquí se tienen en cuenta los municipios con más de 100 mil habitantes.

neoparamilitares[61]. Las restantes ciudades de la Costa configuran un patrón regional cuyo rasgo viene a ser una marcada estabilidad una vez se le coteja con los cambios bruscos del país y Antioquia. Barranquilla oscila entre tasas de 20 y 50 salvo en el pico de 1996, cuando se dispara hasta una anormal tasa de 85 (Gráfico No. 10). Cartagena tiene un ascenso entre 1985 y 1993 pero de allí en adelante permanece estable variando entre tasas de 20 y 40. Entre las ciudades capitales de la región Caribe, Santa Marta es el caso un tanto atípico, en particular por sus tres picos hacia arriba (en 1982, 1992 y 2000-2001) y sus dos picos hacia abajo (en 1983 y 2009), pero de resto permanece con variaciones entre tasas de 40 y 60. Valledupar y Sincelejo son también ciudades representativas del modelo violento de la Costa, oscilan entre magnitudes que van de 20 a 60 con la excepción de un ascenso en Valledupar entre 2000 y 2003[62]. En comparación con el patrón antioqueño, el modelo costeño está marcado por tasas más bajas y por la relativa estabilidad a lo largo del tiempo.

Grafico No. 10
Tasa de homicidio Santa Marta, Barranquilla y Cartagena. 1980-2010

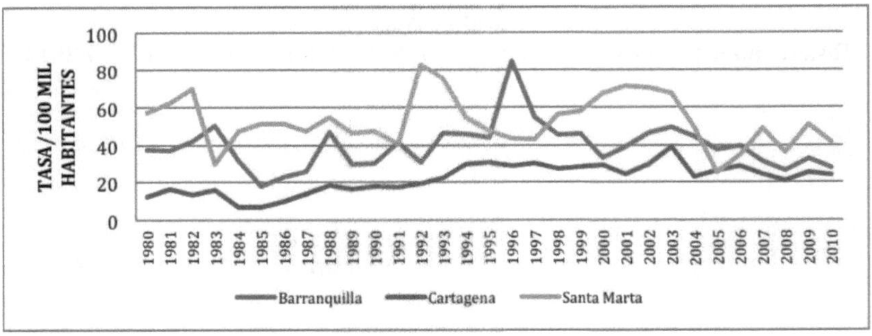

Fuente: DANE. Cálculos nuestros.

El último caso para ejemplificar el comportamiento regional es el del departamento del Valle (Gráfico No. 11). En esta ocasión se sacaron dos curvas, una que reúne las ciudades de Cali, Buga y Buenaventura y otra que condensa las ciudades de Jamundí, Palmira, Tuluá y Yumbo. Otra vez, la coincidencia entre las dos curvas muestra la existencia de un patrón

61 Lorica es igual una ciudad costeña en su comportamiento pero igual con el fuerte cambio iniciado en 2007 por la presencia neoparamilitar. Montería y Lorica pertenecen ambas al departamento de Córdoba.

62 Rioacha y Maicao comienzan los años 80 con tasas elevadas debido a que La Guajira fue el escenario de la fiebre de la marihuana a finales de los 70. En los años que siguen Rioacha pasa a ser la prototípica ciudad costeña moviéndose entre tasas de 20 y 60 con un pico en 2000-2003. Maicao se comporta distinto por su condición de ciudad de frontera.

regional caracterizado por una primera fase de crecimiento sostenido hasta 1994, con una caída hacia finales de la década de los 80; luego viene una segunda fase hasta 1997-1998 que abre paso a un tercer momento de ascenso hasta 2000-2001. La siguiente fase tiene un descenso pero en ningún caso tan pronunciado como lo hace la curva nacional, sino que permanece en tasas altas por encima de 60 con una tendencia al alza durante los últimos años.

Grafico No. 11
Tasa de homicidio de las ciudades del valle. 1980-2010

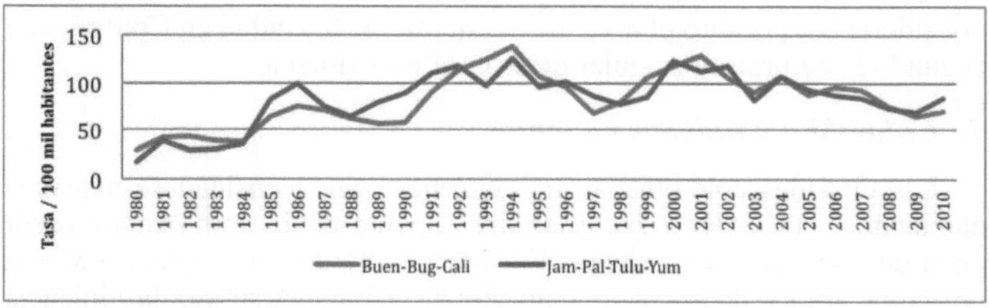

Fuente: DANE. Cálculos nuestros.

La existencia de patrones regionales se resume en el Gráfico No. 12, donde se plasman las curvas que agrupan las tasas de las respectivas ciudades de cuatro regiones. Una vez más las diferencias entre los patrones saltan a la vista. Primero en el comportamiento, luego en la intensidad de las tasas Desbordadas de Antioquia; las Muy Altas y sostenidas del Valle; las Altas y Medias también sostenidas de la Costa Caribe. La ciudad no se contiene a sí misma, hay patrones que dan cuenta de los intercambios con su contexto regional. Las explicaciones del conflicto violento en la ciudad han de ser ligadas al entorno, a los comportamientos y modalidades de acción de los actores de la región en donde ella está inscrita.

Grafico No. 12
Tasa de homicidio por regiones 1980-2010

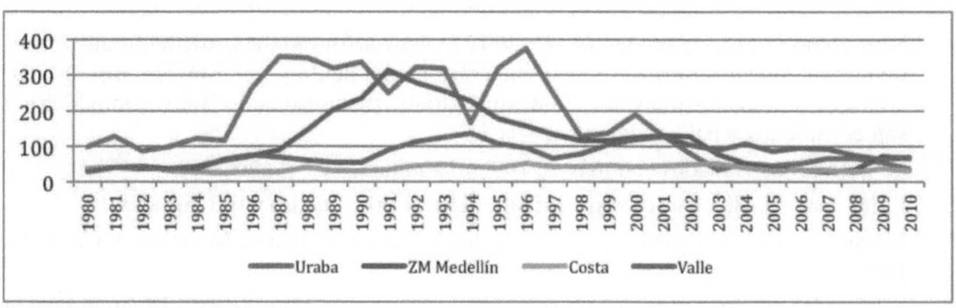

Fuente: DANE. Cálculos nuestros.

III. La singularidad urbana

La noción de "conflicto violento urbano" evoca la existencia de una violencia cuyo rasgo característico pasa no únicamente por la condición de su ocurrencia en la calle de la ciudad, sino también y quizás lo más importante por su conexión con dinámicas de naturaleza propia a los avatares de la vida urbana. La ciudad –lo hemos mostrado– no es el escenario privilegiado del conflicto armado pero mantiene más de una conexión con los conflictos nacional y regional. ¿Qué sucede entonces al interior de la ciudad? ¿Cuál es la naturaleza de su violencia y cómo se conecta con las dinámicas de la guerra? Avanzaremos en la pregunta considerando, primero el nivel de homicidio de la ciudad en Colombia, y segundo la naturaleza singular de su conflicto violento.

1. Colombia: Ciudades violentas

Las ciudades colombianas son violentas. Su intensidad se visualiza atendiendo a un criterio que establece el punto de 25 como cota a partir de la cual se asume la violencia Elevada[63], entre 25 y 11 violencia Media (en color gris) y de 10 para abajo violencia Baja. A su vez la violencia Elevada está dividida entre violencia Desbordada por encima de 100, Muy Alta entre 50 y 99 y Alta entre 26 y 49. En el Cuadro No. 8 aparecen las 60 ciudades de Colombia con sus respectivas tasas de homicidio promedio en cada una de las últimas tres décadas[64]. El primer fenómeno digno de mención es la escasez de violencias Bajas en el Cuadro –escaso color verde–: un mínimo de 6% de los 180 cuadros[65]; asimismo resulta destacable el hecho de que ninguna ciudad se mantenga en nivel Bajo durante las tres décadas, sólo algunas de la Costa lo hacen las dos primeras[66]. El nivel Medio entre 11 y 25 tiene mayor presencia (el color gris suma el 25% de los cuadros), pero sobresalen ante todo los valores de la violencia elevada en sus tres categorías: Alta, Muy Alta y Desbordada, que hacen el 69% de los cuadros. Es el rasgo dominante, durante los últimos 30 años las ciudades colombianas han estado sometidas a violencias elevadas[67]. Nada más 9 se

63 Se escogió la tasa de 25 como cota de la violencia Elevada considerando que Brasil tuvo entre los años de 1996 y 2009 una tasa media de 25 y que México, tras su gran crisis, llegó a ese mismo valor. En el contexto internacional tanto Brasil como México son considerados países violentos.

64 Se trata de las ciudades con más de 100 mil habitantes en el año 2011.

65 El número de 180 cuadros resulta de multiplicar 60 ciudades por 3 décadas.

66 Malambo y Soledad en el departamento del Atlántico; Lorica en Montería; y Uribia en la Guajira.

67 27% de los cuadros en violencia Alta, 28% en violencia Muy Alta y 14% en violencia Desbordada.

mantuvieron las tres décadas con valores medios y bajos, mientras las restantes 51 tuvieron al menos una década con violencia Elevada.

Cuadro No. 8
Tasa promedio de homicidio de las ciudades por décadas. Colombia 1980-2010

CIUDAD	1980	1990	2000	CIUDAD	1980	1990	2000
Medellín	125	267	93	Neiva	23	30	39
Bello	65	184	71	Riohacha	60	43	56
Envigado	56	137	36	Maicao	101	78	85
Itagui	65	218	90	Uribia	3	2	15
Rionegro	58	147	67	Santa Marta	52	55	52
Apartadó	228	316	74	Pitalito	63	90	82
Turbo	156	204	78	Ciénaga	22	51	84
Barranquilla	34	47	38	Villavicencio	51	63	53
Malambo	2	9	20	Pasto	10	25	38
Soledad	0	9	18	Ipiales	25	48	47
Bogotá	18	56	25	Tumaco	19	32	69
Cartagena	13	23	27	Cúcuta	29	102	103
Magangué	10	19	23	Armenia	24	59	57
Tunja	27	32	19	Pereira	69	116	108
Duitama	20	17	19	Dosquebradas	20	34	64
Sogamoso	14	22	16	Bucaramanga	32	59	38
Manizales	43	93	67	Barrancabermeja	30	144	86
Florencia	62	84	90	Floridablanca	6	15	18
Yopal	23	73	72	Girón	18	30	23
Popayán	54	58	56	Piedecuesta	25	30	15
Valledupar	49	39	61	Sincelejo	15	31	27
Quibdó	15	35	54	Ibagué	16	43	31
Montería	25	36	32	Cali	46	104	95
Lorica	2	4	17	Buenaventura	54	100	95
Chía	17	30	15	Buga	55	92	101
Facatativá	19	35	23	Cartago	168	193	122
Fusagasugá	38	44	37	Jamundí	35	103	99
Girardot	22	32	35	Palmira	40	81	81
Soacha	16	25	33	Tuluá	83	126	121
Zipaquirá	22	32	21	Yumbo	63	87	80

Fuente: DANE. Cálculos nuestros.

100 o más	Desbordada
53 a 99	Muy Alta
27 a 52	Alta
11 a 26	Media
10 o menos	Baja

El Cuadro No. 9 ofrece un panorama adicional de los elevados niveles de violencia de la ciudad colombiana. En los 31 años en consideración siete ciudades tuvieron una tasa promedio desbordada, esto es por encima de 100 (rojo); 21 tuvieron tasas Muy Altas, entre 51 y 100 (zapote);

15 tasas Altas entre 26 y 50 (curuba). Tan sólo 17 puntuaron en valores medios y bajos (gris y verde), lo cual significa que el 72% de las ciudades nacionales vivieron los últimos 30 años en violencia elevada. El cuadro contrasta de manera notable con México, donde apenas un 13% de sus ciudades tuvo una tasa promedio elevada entre los años de 1985 y 2010, aunque menos con Brasil, país donde el 53% tuvo un promedio elevado entre los años de 1996 y 2009[68]. La ciudad colombiana está sometida a intensas violencias. Mientras la tasa de sus ciudades con violencia Elevada es de 95, la de Brasil es de 44 y la de México de 40, frente a los dos casos con más de 50 puntos de diferencia[69]. Las ciudades colombianas, no cabe duda, vienen sometidas a crudas violencias.

Cuadro No. 9
Tasa promedio de homicidio de las ciudades. Colombia 1980-2010

CIUDAD	1980-2010	CIUDAD	1980-2010	CIUDAD	1980-2010
Apartadó	206	Manizales	68	Ibagué	30
Medellín	162	Palmira	67	Girardot	30
Cartago	161	Yopal	56	Tunja	26
Turbo	146	Popayán	56	Facatativá	25
Itagui	124	Villavicencio	56	Zipaquirá	25
Tuluá	110	Santa Marta	53	Sincelejo	25
Bello	107	Riohacha	53	Soacha	25
Pereira	97	Ciénaga	53	Pasto	24
Rionegro	91	Valledupar	50	Girón	24
Maicao	88	Armenia	47	Piedecuesta	23
Barrancabermeja	86	Bucaramanga	43	Cartagena	21
Buenaventura	83	Barranquilla	40	Chía	20
Buga	83	Ipiales	40	Duitama	18
Cali	81	Tumaco	40	Sogamoso	17
Jamundí	79	Fusagasugá	40	Magangué	17
Florencia	79	Dosquebradas	39	Floridablanca	13
Pitalito	79	Quibdó	35	Malambo	10
Cúcuta	78	Bogotá	33	Soledad	9
Yumbo	77	Montería	31	Lorica	8
Envigado	76	Neiva	31	Uribia	7

Fuente: DANE. Cálculos nuestros

68 México tiene 175 ciudades de más de 100 mil habitantes, 23 de las cuales tuvieron violencia Elevada. Brasil tiene 273 ciudades, 144 de las cuales tuvieron violencia Elevada.

69 La tasa promedio de las ciudades con violencia Elevada.

Tal magnitud, en un contexto de escasa presencia del conflicto armado, exige claridad sobre la naturaleza del homicidio urbano. Para el efecto apelamos a la información del Instituto de Medicina Legal y Ciencias Forenses entre los años de 2005 y 2011[70]. Se trata de una valiosa información sobre la que no obstante es preciso entrar en toda suerte de matizaciones sobre sus alcances. En realidad –es el punto de partida–, el saber sobre el homicidio en Colombia es fragmentario, poco se esclarece la identidad de los victimarios y las condiciones en que produce sus actos. La disponibilidad de información varía de manera considerable de una a otra variable. El sexo de las víctimas se determina en el 100% de los casos pero de allí en adelante los porcentajes comienzan a disminuir. La edad se establece en el 85% y la escolaridad en el 55%. Los porcentajes disminuyen de manera dramática una vez se entra en el detalle del homicidio: la circunstancia se determina en un 27% y el presunto agresor igual[71]. Por demás el monto de esclarecimiento varía de año a año, a veces con distancias considerables, lo mismo si se mira la ciudad o el poblado. En cada caso se harán las claridades que tales condiciones fuerzan a hacer.

2. *Variables sociodemográficas*

En materia del sexo de las víctimas –la única con información completa todos los años–, no existe diferencia alguna entre el homicidio de la ciudad y el del poblado[72]. Tanto en la una como en el otro (no importa si es la megalópolis o el poblado pequeño), los hombres son asesinados en el 91% o más de los casos y las mujeres en el 9% o menos (Gráfico No. 13). La estabilidad en el tiempo y la constancia sobre la geografía pone al abrigo de sospecha el hecho de que el conflicto violento recae de forma trágica sobre los hombres, una condición que no discrimina entre la calle de la ciudad o la vereda rural.

Grafico No. 13

Sexo de las victimas en ciudades y poblados. Colombia 2005-2011

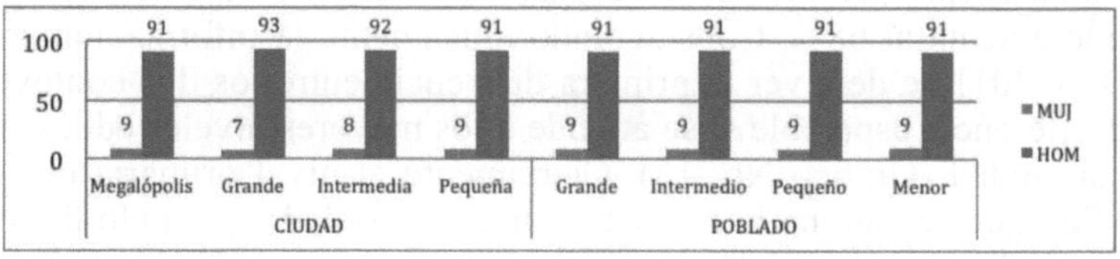

Fuente: Instituto Nacional de Medicina Legal y Ciencias Forenses. Cálculos nuestros.

[70] Nuestra gratitud para Medicina Legal por permitirnos el acceso a esta información desagregada.

[71] Los porcentajes reportados son entre 2005 y 2011.

[72] Hay un debate en marcha en torno a la necesidad de abrir una casilla independiente para los géneros no convencionales, como el transgenerismo.

La edad ya tiene diferencias, en este caso en el tiempo. El sistema de registro de Medicina Legal mejora en la captación de información pues si entre 2005 y 2008 logra determinar la edad en el 73% de los casos, entre 2009 y 2011 logra hacerlo en el 99%. Los datos a tomar en cuenta, así pues, se limitan a los últimos tres años distribuidos en generaciones separadas cada 15 años[73]. Más de la mitad de las muertes violentas son producidas sobre los jóvenes entre 16 y 30 años, seguidas después por los adultos jóvenes entre 31 y 45 (Gráfico No. 14). En tales condiciones las personas entre 16 y 45 años son objeto de más de las cuatro quintas partes de los homicidios. Mas en contravía de lo esperado el asesinato de jóvenes –el que se lleva más de la mitad del total–, no es con preeminencia urbano: la ciudad está apenas 3 puntos porcentuales por encima del poblado. Puesto que la clasificación por generaciones cada 15 años no deja de tener dificultades[74], la reclasificación cada 10 años tampoco arroja una diferencia significativa. Los jóvenes entre 16 y 25 años caen víctimas de homicidio en la ciudad en un 36% y en los poblados en un 33%. En consecuencia la edad tampoco hace la diferencia en el homicidio de la ciudad respecto al del poblado.

Grafico No. 14
Edad de las victimas en ciudades y poblados. Colombia 2009-2011

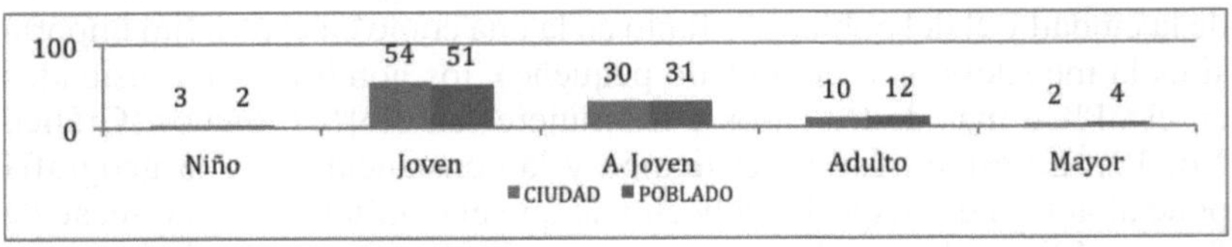

Fuente: Instituto Nacional de Medicina Legal y Ciencias Forenses. Cálculos nuestros.

La última variable sociodemográfica, la escolaridad, sigue el mismo patrón de la edad en tanto la información disponible para los años 2005-2008 da cuenta del 46% de los casos, mientras los siguientes tres años se incrementa al 68%. Considerando únicamente la información entre 2009 y 2011 se deja ver la primera diferencia entre los dos contextos, una diferencia esperable si se atiende a los mayores niveles educativos de la ciudad (Gráfico No. 15). Ciertamente el nivel primaria crece a medida que se camina hacia el extremo del poblado, pasando de 39% en la ciudad grande al 57% en los poblados menores –con la excepción

73 Niños 0 a 15; jóvenes 16 a 30; adultos jóvenes 31 a 45; adultos 46 a 60; y mayores 60 en adelante.

74 La discusión en torno a los rangos de clasificación de lo joven no está zanjada. En México se utiliza el rango de 15 a 29 mientras el último documento oficial de Colombia los coloca entre los 15 y los 25.

de la megalópolis Bogotá–. De contramano el nivel secundaria sigue la dirección opuesta disminuyendo a medida que se avanza hacia el poblado, decreciendo del 49% en la ciudad grande al 24% en el poblado menor (otra vez Bogotá quiebra la tendencia). Los niveles de educación media y superior no manifiestan diferencias, pero no deja de ser cierto que las víctimas de homicidio de la ciudad tienen un nivel de educación más alto.

Grafico No. 15
Escolaridad de las victimas en ciudades y poblados. Colombia 2009-2011

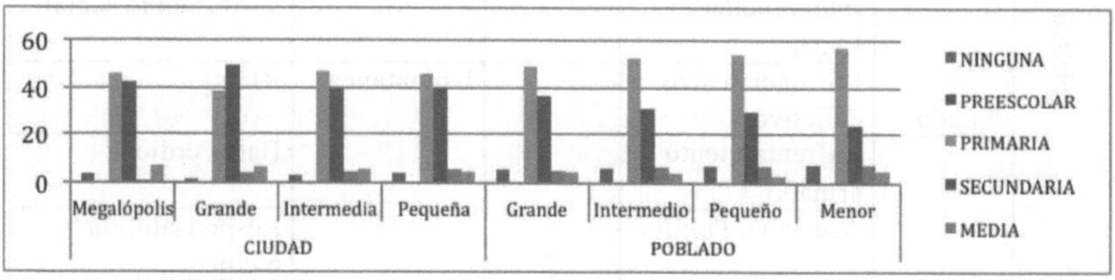

Fuente: Instituto Nacional de Medicina Legal y Ciencias Forenses. Cálculos nuestros.

La gran mayoría de las víctimas tienen un nivel educativo bajo: el 46% tiene primaria y el 41% secundaria, esto es el 87% del total. El porcentaje de Sin Ninguna educación hace un pequeño 4%, los asesinados han ingresado a la escuela. De tal suerte aparece que las víctimas de homicidio en Colombia durante años recientes son abrumadoramente hombres, en su gran mayoría entre los 16 y los 45 años con niveles de educación entre primaria y secundaria.

3. *Las condiciones*

Los datos sociodemográficos muestran cierta consistencia. El salto a las condiciones en las que se produce el homicidio, por desgracia, no viene acompañado de la misma afirmación. Dos variables de la mayor trascendencia para nuestro propósito, la circunstancia y el presunto agresor, reportan información sobre un escaso 27% de los homicidios. La precariedad de esta información obliga a moverse con extrema precaución. En primer lugar y en especial frente a la circunstancia. La categoría Sin Información alcanza su punto más bajo en 2005 con el 66% de los casos (se esclarece el 34%), pero de ahí en adelante oscila entre 70 y 79%. El panorama de la ciudad es más grave, se esclarece nada más el 24% de los casos mientras en el poblado el 32%. Puesto que no se produce una mejoría en los últimos años se emplearán los datos de los cinco años como un modo de compensar las dificultades de información.

Cuadro No. 10
Categorías de la circunstancia

TIPO	ESCE**	EVENTOS	TIPO	ESCE**	EVENTOS
Política	Armados ilegales	Acción Guerrillera	**Social**	Grupos conflicto	Barras, Pandillas, Bandas (2)
		Paramilitar/ Bandas (1)		Identidades	Intolerancia
	Estado	Acción Militar			Contra descalificados o marginales (3)
		Intervención Legal			V-AC* en lo social
		Prevención acto delictivo	**Cotidiana**	Espontánea	Riña
	Eventos	Enfrentamiento armado			Bala Perdida
		Asesinato Político			Responsabilidad médica
		Terrorismo		Intrafamiliar	Violencia de pareja
		V-AC* en lo político			Violencia contra NNA (4)
Económica	Menor	Atraco callejero			Violencia entre otros familiares
		Hurto		Sexual	Delito Sexual
	Mayor	Robo Banco o comercio			V-AC* en lo cotidiano
		Robo Residencia – Morada		Sin información	
		Robo Vehículo	**Sin Informa**	Otros	
		Retención ilegal (Secuestro)		No aplica	
		V-AC* en lo económico			

* V-AC: Venganza – Ajuste de cuentas
** Escenario
(1) Acción paramilitar/Bandas criminales
(2) Conflicto barras, pandillas, bandas
(3) Violencia contra grupos descalificados o marginales
(4) Violencia contra niños, niñas y adolescentes

La variable circunstancia se sometió a un ordenamiento en cuatro tipos[75]: política, económica, social y de la intimidad (Cuadro No. 10). Cada tipo de violencia, a su vez, incluye varios escenarios. Cada escenario, por último,

[75] Las categorías de Medicina Legal se toman tal como ellos las reportan. Aquí se hace simplemente un reordenamiento de ellas en categorías más amplias.

incorpora las clasificaciones empleadas por Medicina Legal en la categoría "circunstancia". Dos clasificaciones de Medicina Legal ofrecen enorme dificultad. Primero la "venganza-ajuste de cuentas" parece funcionar como un comodín en el que cabe cualquier situación. En el intento de precisarla se hizo un cruce con la categoría "presunto agresor" dando lugar a ajustes de cuentas en cada uno de los tipos de circunstancia, un procedimiento que permite un manejo un tanto más discriminado de una categoría que agrupa situaciones dispares. En segundo lugar el "conflicto entre barras, pandillas y bandas" introduce en una sola casilla circunstancias por entero distintas en su naturaleza. De una parte la pandilla es un fenómeno que pasa por la constitución de una identidad local y una apropiación territorial, mientras la banda es una agrupación cuyo propósito expreso es la acumulación de dinero al margen de toda visibilidad local. La barra, entretanto, habla de una agrupación congregada alrededor de una afición deportiva cuyos nexos con la pandilla no son ni mucho menos directos. La barra y la pandilla clasificarían en la circunstancia de tipo social mientras la banda en el tipo económico. Sin embargo es imposible desagregar la categoría, se optó por dejarla en el tipo social.

En esta oportunidad, a diferencia de las variables sociodemográficas, los datos muestran una diferencia que confirma el hallazgo sugerido a propósito de los indicadores del conflicto armado: éste no es el agente dominante de la violencia en la ciudad. El Gráfico No. 16 lo ilustra. En la ciudad la violencia política hace nada más el 11% de los homicidios mientras la violencia cotidiana se lleva el primer lugar con el 28% –en la mitad se sitúan la económica con el 17 y la social con el 11%–. El contraste con el poblado es notorio: la circunstancia de tipo político carga las dos terceras partes de los casos con el 66%, seguido de lejos por la cotidiana con el 15%. Los ajustes de cuentas sin agresor identificado son protuberantes en la ciudad (33%), menos en el campo (11%)[76].

Grafico No. 16
Tipo de circunstancia en ciudades y poblados. 2005-2011

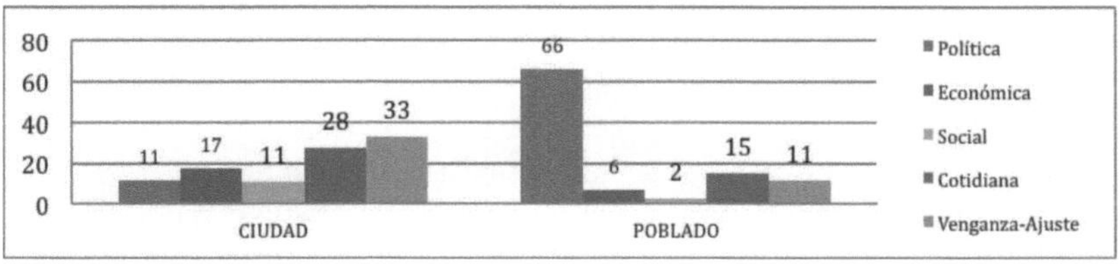

Fuente: Instituto Nacional de Medicina Legal y Ciencias Forenses. Cálculos nuestros.

[76] El dato de venganza-ajuste de cuentas disminuye todavía más la información disponible de la circunstancia.

Por supuesto la prevalencia de una circunstancia no suprime la existencia de las demás. La política no es el tipo de circunstancia predominante en la ciudad, sin embargo del total de 10.769 homicidios asociados a ella –cometidos entre los años de 2005 y 2011–, una cantidad de 2.199 tomaron como escenario la ciudad (Cuadro No. 11). Una suma nada despreciable que alerta frente a las maneras como se dirimen las luchas políticas urbanas. Del mismo modo tampoco son menospreciables las violencias económica, social y de la intimidad en los poblados: de los 15.105 asesinatos asociados al tipo social la considerable cantidad de 3.380 sucedieron en ellos.

Cuadro No. 11
Numero de homicidios según tipo de circunstancia. 2005-2011

	CIUDAD	POBLADO	TOTAL
POLÍTICA	2.199	8.570	10.769
ECONÓMICA	2.742	812	3.554
SOCIAL	11.725	3.380	15.105
INTIMIDAD	761	384	1.145
TOTAL	17.427	13.146	30.573

Fuente: Instituto Nacional de Medicina Legal y Ciencias Forenses. Cálculos nuestros.

El cuadro desagregado por escenarios termina de clarificar la diferencia entre ciudad y poblado (Gráfico No. 17)[77]. En este último se producen las dos terceras partes y más de los homicidios en los cuatro escenarios de tipo político: el 85% por grupos armados ilegales, el 81% por el Estado y el 79% por eventos del conflicto armado. La situación opuesta se verifica en los otros tres tipos, donde la ciudad se pone por encima en todos los escenarios. En el económico y social provoca más del 70% de los homicidios en los cuatro escenarios. Finalmente, con una diferencia menor la ciudad también domina los tres escenarios de la circunstancia de tipo cotidiano, un poco menos en el homicidio asociado a lo sexual.

[77] En total en el poblado se cometen el 80% de los homicidios políticos; en la ciudad el 77% de los económicos, el 78% de los sociales y el 66% de la intimidad.

Grafico No. 17
Escenarios de la circunstancia en ciudad y poblado. 2005-2011

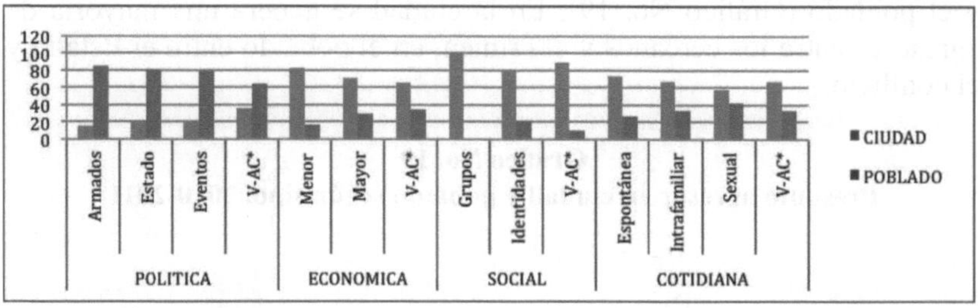

* Venganza – ajuste de cuentas[78]
Fuente: Instituto Nacional de Medicina Legal y Ciencias Forenses. Cálculos nuestros.

La siguiente variable de las condiciones en las que se produce el homicidio se refiere al perpetrador, uno de los indicadores de mayor valor en la tarea de su esclarecimiento. Afortunadamente el acopio de su información mejora en los dos últimos años pasando de 21 a 44%[79]. Como en el caso de la circunstancia el agresor ha sido clasificado en cuatro tipos: cercano; crimen y seguridad; estado; y conflicto armado. Cada tipo está a su vez clasificado en actores[80] y cada uno de ellos, del mismo modo, contiene el victimario tal y como la reporta Medicina Legal. Partiendo de tal clasificación el presunto agresor confirma la situación de la ciudad. En ella prevalece la violencia proveniente de personas cercanas (51%), seguida de la criminalidad (32%), sumando entre los dos el 83% de los homicidios urbanos –los otros dos, Estado y conflicto, suman un pequeño 17%– (Gráfico No. 18); en contraste en el poblado los tipos Estado y conflicto agregan el 65% de los homicidios, mientras cercanos y crimen el 35%.

Grafico No. 18
Presunto agresor por tipo en ciudad y poblado. 2010-2011

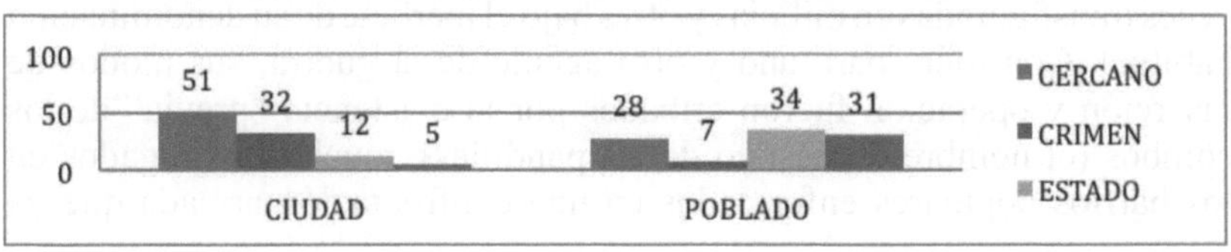

Fuente: Instituto Nacional de Medicina Legal y Ciencias Forenses. Cálculos nuestros.

---

78 Como se anotó la venganza-ajusta de cuentas se distribuyó en los 4 tipos según el presunto agresor.

79 Se trabajará entonces con la información de los años 2010 y 2011.

80 Cercano: familiar, pareja, amigo, contractual y conocido sin ningún trato/ Crimen y seguridad: bandas criminales, delincuencia, pandilla, narcotráfico y seguridad privada/ Estado: fuerzas armadas y seguridad/ Conflicto armado: guerrilla y paramilitares.

La misma situación pero vista desde la distribución de cada tipo de agresor ratifica la abismal diferencia entre las violencias de la ciudad y el poblado (Gráfico No. 19). En la ciudad se genera una mayoría de agresores entre los cercanos y el crimen, en el poblado entre el Estado y el conflicto.

Grafico No. 19
Presunto agresor en ciudad y poblado según tipo. 2010-2011

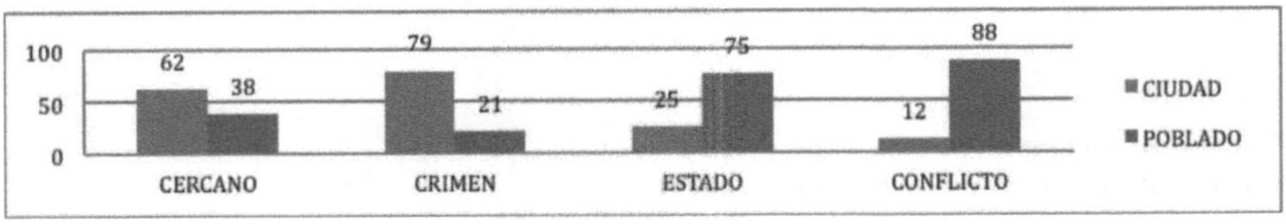

Fuente: Instituto Nacional de Medicina Legal y Ciencias Forenses. Cálculos nuestros.

La ciudad posee una singularidad, una condición particular asociada a las gramáticas de la vida privada y la criminalidad; el poblado, por el contrario, liga el ejercicio de su violencia al contexto del conflicto armado. El trabajo etnográfico lo confirma, los actores armados de la guerra no son, ni muchos menos, el eje del conflicto urbano y su resolución sangrienta[81]. Claro, la guerra y sus incidencias aterrizan en la calle –lo mostramos páginas atrás–, pero lo hacen trabando un complejo intercambio con las dinámicas propias de la ciudad.

Medellín, sede de astronómicas violencias –la única ciudad grande donde el conflicto armado ha llegado al punto de inducir desplazamientos masivos de zonas enteras[82]–, revela la anudada trama entre singularidad urbana y guerra. La fuerza paramilitar no tuvo que ser exportada al estilo de un frente declaradamente interesado en tomarse la ciudad[83], ella tenía asiento de tiempo atrás como resultado de la prolongada historia de sus mafias internas. Las guerrillas también hicieron su ingreso triunfal, unas veces transfiguradas en milicias y otras bajo el marbete de su denominación habitual. Con todo, para uno y otro agente de la guerra, sus modos de inserción y operación fueron cribados por la existencia "previa" de los combos (el nombre doméstico de las pandillas), muchachos regados en los barrios populares enfrascados en una confrontación armada que ya

[81] Nos hemos centrado en información numérica con el propósito de construir un marco general a la investigación de la ciudad. En las seis ciudades en comparación, sin embargo, se avanza un intenso trabajo etnográfico orientado al rastreo del conflicto violento en cada una de ellas.

[82] Medellín, con sus 2.368.282 habitantes en 2011, ingresa en la categoría de ciudad grande.

[83] Como fuera el caso de Bogotá, "tomada" por los frentes Capital y Casanare.

bordea los 30 años de permanencia[84]. Un fenómeno paradigmático del contexto urbano –la pandilla–, opera como clave de anudamiento de la intensa y prolongada violencia de Medellín. Luego de la operación militar del ejército en 2002 el poder mafioso paramilitar se tomó a sus anchas la ciudad, una vez el Estado les terminara de librar de la insurgencia. De en ese momento en adelante el poder centralizado en un solo "señor" (Don Berna), significó el control de bandas cuyo poderío, a su vez, descansa sobre la afiliación de los combos diseminados en las barriadas de la periferia[85]. La guerra se procesa en la singularidad de la ciudad; por supuesto, en el camino de vuelta, la frenética agresividad de los combos paisas proviene de su conexión directa con la guerra.

Para concluir

La separación de las violencias hace carrera en la reflexión internacional. El texto *Carga Global de la Violencia Armada* introduce una diferenciación de gran pertinencia para el punto en consideración (Geneva Declaration, 2008). Existen tres clases de violencia –dice el informe–: aquellas producidas de manera directa en un conflicto, producto de los enfrentamientos entre las fuerzas en choque, incluyendo soldados y civiles; otras provocadas de forma indirecta por un conflicto, conectadas a las enfermedades, el hambre y la desnutrición ocasionadas por la destrucción de las condiciones de vida durante la guerra; y una más articuladas a violencias que no guardan relación alguna con un conflicto armado. La clasificación se justifica una vez se contempla el panorama mundial. Entre los años 2004 y 2007 murieron 740 mil personas en situaciones de violencia letal, distribuidas entre un reducido 7% en contextos de conflicto armado (directas), un 27% por la destrucción de la guerra (indirectas) y un 66% en situaciones ajenas a la confrontación bélica (no asociadas)[86]. El panorama mundial se reproduce en Colombia –aún en medio de un agudo conflicto armado–, según lo confirma la baja participación del homicidio político en el conjunto del homicidio nacional (González, Bolívar y Vázquez, 2006; Gutiérrez, 2006)[87].

---

84 Una comparación de Medellín y Río de Janeiro en Perea (2014a).

85 Mientras se escriben estas páginas Medellín afronta una amarga disputa por el control de la cúspide, partida ahora entre lo que queda de la oficina de Envigado y los Urabeños –aunque en julio de 2013 se trazó un pacto que hasta hoy se mantiene–.

86 El informe muestra la multiplicidad de violencias pero sigue ligando el conflicto a la guerra. Optamos así por la noción de conflicto violento como un universal abierto en diversas violencias.

87 En su pico más elevado la violencia política produjo el 25% del homicidio del país, pero en otros momentos contribuye con menos del 10% de las muertes.

De tal suerte, y recogiendo la reflexión de estas páginas, se perfilan dos circunstancias críticas frente a la paz. De un lado una enorme cantidad de homicidios cometidos fuera de la guerra; del otro la ciudad como el gran escenario donde toman cuerpo esas muertes no conexas al conflicto armado. ¿Qué lecciones ha de tomar entonces la paz de la ciudad? La abigarrada agencia de la violencia urbana supone la definición de una política de paz más allá de las negociaciones con las guerrillas. En efecto, el abanico de agentes del homicidio en la ciudad es más que amplio: crimen organizado, delincuencia común, pandillas, grupos de seguridad vecinales, muertes espontáneas (ligadas a riñas), cercanos (repartidos entre familiares, amigos, vecinos y próximos por alguna relación) y operaciones de limpieza (las que jamás aparecen en los registros y discursos oficiales). Un amplio espectro tras del cual surge la importancia de definir políticas sectoriales ligadas al proyecto nacional de paz.

La violencia cotidiana –producida por personas cercanas–, dotada en los registros de considerable presencia (de seguro sobre registrada dada la comparativa facilidad de establecer su circunstancia y agresor), pone al aviso de los niveles de penetración de la "solución" violenta en los intercambios de la vida diaria. El tema de la convivencia salta a la palestra de las urgencias de la paz. En un país arrastrado por un largo e intenso conflicto violento –marcado por un penetrante homicidio urbano–, la reconstrucción de los tejidos sociales se perfila como condición primordial de las condiciones que garanticen el establecimiento de una paz estable y duradera.

Otro tanto hace la violencia económica. Sus indicadores no dejan de ser altos –y con seguridad sometidos a un elevado subregistro–, dando cuenta de una violencia que se libra en el contexto de una nación y unas ciudades, por un lado cruzadas por una de las inequidades más flagrantes del planeta y por otro invadidas por toda suerte de circuitos mafiosos e ilegales. En el cruce de caminos de la inequidad y la ilegalidad aparece presta la violencia. En dirección por completo opuesta a la sostenida tesis de la disociación entre la pobreza y la violencia –según la cual la violencia en Colombia se dispara en los lugares de las grandes rentas[88]–, las ciudades con sus elevadas violencias evidencian que la muerte se implanta allí donde campean la pobreza y la marginalidad[89]. Es el otro franco desafío que lanza la ciudad a la paz: el horizonte de la equidad es la otra condición esencial de una paz que pretende ser sólida y permanente en el tiempo.

El conflicto violento no se agota en el conflicto armado. Perseverar en el achatamiento supone enfrentar la paz ateniéndose a una lógica

---

88 Fue la posición de los citados economistas que luego el uribismo llevó hasta sus últimas consecuencias.

89 En la ciudad el mapa de la violencia coincide con el mapa de la marginalidad.

que visibiliza e invisibiliza violencias según el nexo que ellas guardan con la guerra. La operación de limpieza, una violencia de naturaleza social, ilustra bien el punto desde el extremo de una violencia urbana considerablemente olvidada; pese a ser un actor permanente en varias ciudades, ni siquiera se le menciona hasta cuando los actores armados la emplean, como aconteció con las incursiones urbanas de los paramilitares a comienzos de la década anterior[90].

Resituar la ciudad: frente a la reducción mostramos que la ciudad tiene su singularidad –no se agota en el conflicto armado–, pero mostramos también que tampoco es autónoma respecto al curso de la guerra. El diálogo y la concertación con los alzados en armas es el primer paso (obligado) del proceso de la paz; mas el conflicto violento va más allá de la guerra, en particular hacia un contexto urbano con mayor presencia de violencias asociadas a situaciones de otra naturaleza. Guatemala y El Salvador produjeron violencias más intensas y fragmentadas una vez consolidaron sus acuerdos de paz; Colombia posee ya una violencia fragmentada e intensa en las ciudades que la paz debe voltear a mirar con renovados ojos. El país, imbuido hoy en el logro de la paz, dirigirá un considerable monto de sus energías hacia los sectores rurales como consecuencia del peso histórico de varios factores: el conflicto agrario nunca resuelto y la tantas veces demandada reforma agraria; la acumulada deuda histórica con las víctimas, en su mayoría pertenecientes al campo –en la ley las víctimas se definen de nuevo como "víctimas del conflicto armado"–; la política de restitución de tierras; la desmovilización de los combatientes de las FARC con gran fuerza hacia zonas campesinas. El país se volcará sobre el agro. La ciudad vuelve y queda en un limbo y de allí la pertinencia de estas páginas.

La ciudad le habla a la paz desde la inequidad y la convivencia rota. De todo eso trata pues el empeño de resituar la ciudad: de urbanizar la conciencia pública del conflicto, de politizar la mirada sobre la ciudad.

Bibliografía

Avila, Ariel y Nuñez, Magda (2009). "Bogotá cercada, neoparamilitarismo y bandas". En: *¿El declive de la seguridad democrática?* Corporacion Nuevo Arco Iris: Bogotá.

Azaola, Elena (2012). "La violencia de hoy, las violencias de siempre". En: *Desacatos. Revista de Antropología Social.* Centro de Investigaciones y Estudios Superiores en Antropología Social (CIESAS): Ciudad de México.

90 De las operaciones de limpieza no se habla ni siquiera hoy día, pero tuvieron publicidad cuando los paramilitares las emplearon en Altos de Cazucá en la frontera de Bogotá con Soacha. La revista Semana (2003) reportó estos sucesos.

Bedoya, Diego y Jaramillo, Julio (1991). *De la barra a la banda*. El propio bolsillo: Medellín.

Blair, Elsa (2004). *Muertes violentas. La teatralización del exceso*. Instituto de Estudios Regionales (INER). Universidad de Antioquia: Medellín.

Camacho, Alvaro y Guzmán, Alvaro (1990). *Colombia. Ciudad y violencia*. Foro por Colombia: Bogotá.

Centro Internacional para la Prevención de la Criminalidad (2010). *Prevención del crimen y seguridad comunitaria*. En línea.

Comisión de Estudio de la Violencia (1987). *Colombia: Violencia y democracia*. Colciencias-Universidad Nacional de Colombia: Bogotá.

Cubides, Fernando; Olaya, Ana y Ortiz, Carlos Miguel (1998). *La violencia y el municipio colombiano. 1980-1997*. Centro de Estudios Sociales de la Universidad Nacional de Colombia: Bogotá.

DANE (1980-2010). *Estadísticas Vitales. Defunciones por causa externa (1980-2010)*. Cuadro No. 4A.

DANE (2005). "Conceptos básicos". En: *Censo General 2005*. <http://www.dane.gov.co/files/inf_geo/4Ge_ConceptosBasicos.pdf>.

Daza, Fernando (1995). "Una indagación sobre las causas de la violencia en Colombia". En: *Dos ensayos especulativos sobre la violencia en Colombia*. Fonade – Departamento Nacional de Planeación: Bogotá.

Duncan, Gustavo (2005). *Del campo a la ciudad: la infiltración urbana de los señores de la guerra*. Universidad de los Andes. Documento CEDE 2005-2: Bogotá.

Dunca, Gustavo (2006). *Los señores de la guerra. De paramilitares, mafiosos y autodefensas en Colombia*. Planeta: Bogotá.

Echandía, Camilo (1999). *El conflicto armado y las manifestaciones de violencia en las regiones de Colombia*. Presidencia de la República. Observatorio de Violencia: Bogotá.

Espinal, Manuel; Giraldo, Jorge y Sierra, Diego (2007). "Medellín: el complejo camino de la competencia armada". En: *Parapolítica. La ruta de la expansión paramilitar y los acuerdos políticos*. Corporación Nuevo Arco Iris – Intermedio: Bogotá.

Ferro, Juan y Uribe, Graciela (2002). *El orden de la guerra. Las FARC-EP: Entre la organización y la política*. Centro Editorial Javeriano (CEJA): Bogotá.

Flacso Chile (2010). *Reportes del sector seguridad en América Latina y el Caribe*. Disco interactivo.

Geneva Declaration (2008). *Global Burden of Armed Violence*. Geneva Declaration Secretariat: Ginebra.

Giraldo, Jorge (2011). "Economía criminal en Antioquia y Valle de Aburrá". En: Jorge Giraldo (editor). *Economía criminal en Antioquia: Narcotráfico*. Universidad Eafit-ProAntioquia-Empresa de Seguridad Urbana: Medellín.

Gómez, Hernando (2003). *Conflicto: callejón con salida*. Informe de Desarrollo Humano. Organización de las Naciones Unidas: Bogotá.

González, Fernán; Bolívar, Ingrid y Vázquez, Teófilo (2006). *Violencia política en Colombia. De la nación fragmentada a la construcción del estado*. Cinep: Bogotá.

Grupo de Investigación Casa de Paz (2009). "Medellín, la inseguridad cíclica y la permanencia de la ilegalidad". En: *¿El declive de la seguridad democrática?* Corporacion Nuevo Arco Iris: Bogotá.

Gutiérrez, Francisco; Wills, María Ema y Sánchez, Gonzalo (editores) (2006). *Nuestra guerra sin nombre. Transformaciones del conflicto en Colombia*. Instituto de Estudios Políticos y Relaciones Internacionales (IEPRI)-Universidad Nacional de Colombia-Norma: Bogotá.

Gutiérrez, Francisco (2006). "Tendencias del homicidio político en Colombia. 1975-2004". En: *Nuestra guerra sin nombre. Transformaciones del conflicto en Colombia*. (IEPRI)-Universidad Nacional de Colombia-Norma: Bogotá.

Instituto Nacional de Medicina Legal y Ciencias Forenses (2011). *Forensis*: Bogotá.

Instituto Nacional de Medicina Legal y Ciencias Forenses (2000-2008). *Forensis*. Bogotá.

IBGE (1996-2009). MS/SVS/DASIS – Sistema de Informações sobre Mortalidade – SIM. Mortalidade por Agressão no Brasil: 1996-2009.

Jaramillo, Ana María (2011). "Acerca de los estudios sobre conflicto armado y violencia urbana en Medellín". En: Jorge Giraldo (editor). *Economía criminal en Antioquia: Narcotráfico*. Universidad Eafit-ProAntioquia-Empresa de Seguridad Urbana: Medellín.

Jaramillo, Ana María; Ceballos, Ramiro y Villa, Marta Inés (1998). *En la encrucijada. Conflicto y cultura política en el Medellín de los noventa*. Alcaldía de Medellín-Programa para la Reinserción-Región: Medellín.

Leal, Francisco (editor) (1996). *Tras las huellas de la crisis política*. Instituto de Estudios Políticos y Relaciones Internacionales (IEPRI). Universidad Nacional de Colombia-Tercer Mundo-Fescol: Bogotá.

Leal, Francisco (editor) (1995). *En busca de la estabilidad perdida. Actores políticos y sociales en los años noventa*. Instituto de Estudios Políticos y Relaciones Internacionales (IEPRI). Universidad Nacional de Colombia-Tercer Mundo-Colciencias: Bogotá.

Leal, Francisco y Zamosc, León (editores) (1990). *Al filo del caos. Crisis política en la Colombia de los años 80*. Instituto de Estudios Políticos y Relaciones Internacionales (IEPRI). Universidad Nacional de Colombia-Tercer Mundo: Bogotá.

Montenegro, Armando y Posada, Carlos (1995). "Criminalidad en Colombia". En: *Coyuntura económica*. Bogotá, Vol. 25, No. 1.

Noche y Niebla (2008). Marco Conceptual. Banco de datos de derechos humanos y violencia política. Cinep: <http://www.nocheyniebla.org/files/u1/comun/marcoteorico.pdf>.

Organización Panamericana de la Salud – Organización Mundial de la Salud (2003). *Informe Mundial sobre la violencia y la salud*. Washington.

PNUD: Programa de las Naciones Unidas para el Desarrollo (2011). *Informe Nacional de Desarrollo Humano 2011. Colombia rural, razones para la esperanza*. Bogotá.

Policía Nacional (1990-2010). *Revista Criminalidad*. Bogotá.

Pécaut, Daniel (1987). *Orden y violencia: Colombia 1930-1953*. Cerec-Siglo XXI: Bogotá, II tomos.

Pécaut, Daniel (2003). *Guerra contra la sociedad*. Espasa: Bogotá.

Perea, Carlos Mario (2014a). "La muerte próxima: vida y dominación violenta en Río de Janeiro y Medellín". En: *Análisis Político*. Instituto de Estudios Políticos y Relaciones Internacionales (IEPRI). Universidad Nacional de Colombia. No. 80, enero-abril de 2014.

Perea, Carlos Mario y Rincón, Andrés (2014). "Robar pero no matar. Crimen, homicidio y violencia en Bogotá". En: *Ciudades en la encrucijada. Conflicto violento y poder criminal en Río de Janeiro, Medellín, Bogotá y Ciudad Juárez*". Ana María Jaramillo y Carlos Mario Perea (Editores). En prensa.

Perea, Carlos Mario (2008). *¿Qué nos une?* Instituto de Estudios Políticos y Relaciones Internacionales (IEPRI). Universidad Nacional de Colombia – La Carreta Editores: Medellín.

Perea, Carlos Mario (2007). *Con el diablo adentro. Pandillas, tiempo paralelo y poder*. Siglo XXI: México.

Pérez, Bernardo (2007). "El paramilitarismo en Cundinamarca y Bogotá". En: *Parapolítica. La ruta de la expansión paramilitar y los acuerdos políticos*. Corporación Nuevo Arco Iris – Intermedio: Bogotá.

Restrepo, Jorge y Aponte, David (editores) (2009). *Guerra y violencias en Colombia. Herramientas e interpretaciones*. Centro de Recursos para el Análisis del Conflicto (CERAC)-GTZ-Universidad Javeriana-ODECOFI-Colciencias: Bogotá.

Restrepo, Juan (2009). "Estructura paramilitares desmovilizadas en Medellín". En: 2009. *¿El declive de la seguridad democrática?* Corporacion Nuevo Arco Iris: Bogotá.

Rubio, Mauricio (1999). *Crimen e impunidad. Precisiones sobre la violencia*. Tercer Mundo-Centro de Estudios para el Desarrollo (CEDE) – Universidad de los Andes: Bogotá.

Salazar, Alonso (1990). *No nacimos pa'semilla*. Región-Cinep: Bogotá.

Salazar, Alonso y Ana María Jaramillo (1992). *Medellín. Las subculturas del narcotráfico*. Cinep: Bogotá.

Sánchez, Gonzalo (2007). "Los estudios sobre la violencia: balance y perspectivas". En: Gonzalo Sánchez y Ricardo Peñaranda (compiladores). *Pasado y presente de la violencia en Colombia.* Instituto de Estudios Políticos y Relaciones Internacionales (IEPRI). Universidad Nacional de Colombia – La Carreta Editores: Medellín.

Secretaría de Salud de México (1990-2010). *Sistema de Información en Salud.*

Semana (2003). "Masacre gota a gota". Número 1115, 15 de septiembre.

United Nations Organization (2011a). "Urban population 1950-2050". En: *World Urbanization Prospects: The 2011 Revision.* Department of Economic and Social Affairs/Population Division.

United Nations Organization (2011b). "Methodology". En: *World Urbanization Prospects: The 2011 Revision.* Department of Economic and Social Affairs/ Population Division.

From the Desire to Live to the Readiness to Kill

The Result of Legal and Extra-legal Punitive Practices Involving Children and Adolescents in Rio de Janeiro

Joana Domingues Vargas & Natasha Elbas Neri

In recent decades Brazil has experienced high homicide rates. The total number of homicides between 1979, when the Ministry of Health started to monitor the homicide rate, and 2009, the thirtieth year for which records had been kept, reached the startling figure of one million (Zilli & Vargas, 2013). In most of these homicides the victims or perpetrators were black or mixed-race youths and adolescents living in *favelas* (slum communities) or on the outskirts of cities. The victims were murdered because of their involvement in the drug trade, in gangland disputes over territory or in other conflicts involving firearms. Some of these deaths were the result of confrontations or alleged confrontations between youths and the police.

Police lethality in Brazil is extremely high. According to the Brazilian Yearbook of Public Safety, which was produced jointly with the Open Society Foundation, at least five people died every day as a result of police activities in 2012. The study found that the Brazilian police kill more people than the police in countries with high crime rates, such as Mexico, South Africa and Venezuela. Comparison of the data on police lethality in New York, São Paulo and Rio de Janeiro reveals the extent of the problem of police violence in Brazil. In New York, a city with around 8.2 million inhabitants, the police shot 24 people and killed 8 in 2011. In the city of São Paulo, which has a population of eleven million, the police killed 242 people in the same year, while in Rio de Janeiro, which has a population of six million, the corresponding figure was 283 (Anuário Brasileiro de Segurança Pública, 2013).

The increase in homicides among youths and police violence involving young people living in favelas, whether they are involved in drug trafficking or not, is particularly visible in Rio de Janeiro, where a form of micro-trafficking linked with the control of territory developed in favelas and other poor areas, leading to a systematic increase in violence from the 1980s until the high rate of seventy homicides per hundred thousand inhabitants was reached in the mid-1990s. In turn, police measures to tackle drug trafficking not only became more common but were also intensified, reaching a peak in 2007, when 902 civilian deaths in confrontations or alleged confrontations with the police were recorded (Misse *et al.*, 2013).

The situation in Rio de Janeiro has now changed. Starting in December 2008, in preparation for the international events the city was to host in the following years, an ambitious policing policy underscored by proximity with the community was implemented with the objective of "restoring" territories occupied by criminal groups and of "pacifying" the hill sides (*morros*) and favelas. The Pacifying Police Units (*Unidades de Polícia Pacificadora*-UPP) were implemented in tandem with the social policies and plans to reintegrate controlled areas into the city. To date, UPPs have been set up in thirty-eight favelas formerly under the control of armed traffickers, but much work remains to be done before they are extended to the one thousand favelas in Rio de Janeiro (Leite, 2012). Regardless of the explicit or implicit objectives of UPPs, it is a fact that they represent progress compared with previous police practices, which were based only on violent confrontation.

This article addresses the period that precedes this transformation and does not attempt to determine to what extent the scenario described has been effectively modified by this new reality. It seeks to understand how and why, despite the adoption of the principles of childhood protection in Brazil, severe punitive practices still prevail and have in fact been intensified among children and teenagers from lower classes; how the police have become and remained the key agent dealing with this population; and what juvenile delinquents' perceptions of policemen and their actions are and how they react to them.

Although Brazilian legislation on childhood is effective nationwide, the city of Rio de Janeiro, the country's capital until 1960, was chosen as the focus of this study. As the state capital and former national capital, this city has had an important role in changes in legislation related to childhood and youth. Moreover, Rio de Janeiro has been transformed into the country's showcase by the media, and for this very reason what happens in this city exerts enormous influence on domestic behaviors and habits. In Rio, the legacy of the past – the perpetuation of certain punitive practices – and the even more violent new legal and extra-legal forms of conflict resolution are evident for all to see and stand out against practices in the rest of the country. As already mentioned, the city of Rio de Janeiro has one of the worst police lethality indexes in Brazil.

Our analysis of teenagers' perceptions of their experiences with the police is based on a study conducted in two juvenile detention centers in

Rio de Janeiro housing teenage offenders[1] who had been ordered by courts to undergo "socioeducational detention in a juvenile facility."[2] The first detention facility, which we will call the *House*, is meant for teenagers who committed offenses at the ages of 16 or 17. The second, which we shall call the *Little House*, shelters delinquent juveniles who were caught when they were between the ages of 12 and 15. Informal semi-structured interviews combined with continuous observation of routine activities encouraged subjects to speak freely about their experiences with the police and what they referred to as their "life in crime." In addition to the interviews, a questionnaire was given to a sample of 105 inmates in the *Little House*,[3] the only detention center that approved our request to proceed with this kind of investigation.

A brief history of official policies for child offenders

According to the 1890 Republican Code, when youths over the age of nine years were punished they had to be sent to correctional facilities for minors. However, as such institutions were almost non-existent, these offenders ended up incarcerated in common prisons together with adults. This situation was denounced by legal experts, doctors, and those working with childhood assistance, who were startled by scenes of children in the midst of adults who had been convicted as criminals (Rizzini, 1995).

Such scenes were undoubtedly affected by enforcement of the "law to repress idleness", which had been approved less than one year after the abolition of slavery in 1888. The aim of the law, which was the result of parliamentary debates in which former slaves were characterized as lazy vagabonds and addicts prone to commit crimes, was to control the lower classes and, using force, oblige them to work. Hence, Brazil's poor urban black population was often depicted as belonging to dangerous classes (Chalhoub, 1990) and was subjected to disciplinary action by the police, who could act based on the artifice of any "suspicious attitude" (Chalhoub, 1990; Batista, 1998). At that time, the actual administration

1 In Brazil, minors between the ages of 12 and 17 who commit offenses are judged by a special judge at the Children's and Adolescents' Court, according to a specific statute, the Children's and Adolescent's Statute. Unlike in other countries, in Brazil offenses by adolescents are not considered crimes and are not included in their criminal record. If convicted, they do not face judicial sentences but rather must undergo socioeducational measures that can range from community service to detainment at a juvenile detention center, where they can be held for up to three years or until they reach the age of twenty-one.

2 Ethnographic data used in this article were produced during field research conducted in 2007 and 2008 by Neri (2009). All the teenagers' names have been changed.

3 At that time, *Little House* had around 130 inmates, but not all of them were willing to participate.

of *minors* and the final decision on their fate were the responsibility of the police. Detaining young people, classifying their behavior and deciding what should be done with those who were detained were thus police duties. As a result, this unprivileged section of the population became legitimate objects of police action (Vianna, 1999).

As part of the far-reaching movement for the reform of public policies aimed at children in the United States, Europe and Latin America, Brazil made the transition during the first two decades of the 20th century from a repressive penal model to a more welfare-based, protective one that paved the way for the 1927 Minors' Code. After this reform, the police lost partial control over *minors* when a Minors' Court was created in the Federal District in 1923. However, they remained responsible for the first contact with minors and for information about them. In addition, they remained directly and indirectly involved in the organization of detention facilities, long considered part of the solution for poor youths, which basically meant removing them from the streets and educating them for work.

The Minors' Code in turn adopted a classification system similar to that developed by the police (Vianna, 1999). The classifications, which were based on a stigmatizing scale, began with the "materially abandoned", who were followed by the "morally abandoned, "beggars and vagabonds" and, finally, "delinquents" depending on how serious their misdeeds were. These classifications remained officially in effect until 1979, when the new Minors' Code was enacted. Similarly, the practice of diagnosing *minors* continued to be common among court officials and experts as a tool for assessing profiles and monitoring the reeducation of young offenders. This technique, so brilliantly described by Michel Foucault in *Discipline and Punish*, is still employed as a means of social reintegration, especially in detention centers.

The history of Brazilian childhood legislation and practices has been marked by successes and failures (Rizzini, 2000; Sposato, 2006; Vargas & Marinho, 2008). Despite, or perhaps because of this erratic progress, several key issues remain to this day: the criminalization of youths based on their color (black) or financial situation (poor) (Alvim & Valladares, 1988; Batista, 1998); the prominent role of the police in handling this population (Vianna, 1999); the educational guidance offered to these young people to help them achieve nothing more than subaltern jobs (Rizzini, 1995; Alvim, 1995); and the prioritization of investments in confinement measures, whose duration is usually undefined (Batista, 1998).

From a broader perspective, the following elements also constitute part of this legacy: the violent and arbitrary "techniques" employed by the police force in their approach and in their attempts to control populations whose heritage is the stigma of slavery; the associating of these populations with dangerous classes, albeit under new representations; and, finally, the wide gap between legal provisions and actual practices, which is undoubtedly a consequence of severe and persistent social inequalities and the practice in the legal system of treating unequal people unequally (Lima, 1989).

As argued in the literature on this topic, law and justice systems intended specifically for *minors* in Brazil were created to control poor black children and not meant for children in general. And once legal practices were directed at poor black youths, the category *minor* became associated with this section of the population. Formerly a legal definition, the term *minor* was turned into an attribute suggestive of delinquency, crime, poverty, and skin color (Alvim & Valladares, 1988). Police practice served to reinforce and consolidate this specific connotation (Vianna, 1999).

It is for this reason that the Children's and Adolescent's Statute (ECA) rejected the category *minor* and adopted the terms children and adolescents instead. This new legislation was enacted in 1990 in response to punitive legal and extra-legal measures (including mandatory confinement and action by extermination groups, respectively) taken against children and teenagers. However, despite having vanished from official discourse, the term *minor* has endured in police jargon, in certain segments of society (Alvim, 1995), and even among youths who are classified as such (Neri, 2009).

Regarding practices in the juvenile justice system, since the advent of the ECA, juvenile judges have lost some of their power, while the Public Prosecutor's Office has seen its power increase. This suggests yet another shift in terms of who is in charge of decision-making related to young offenders in Brazil: from the police to the juvenile court and its specialists, and from these to the Public Prosecutor's Office. However, the extent to which the latter is currently responsible for initial screening in the juvenile justice system (Silva, 2010) has yet to be investigated. The findings of a study conducted at the juvenile court in Rio de Janeiro suggest that police officers are usually the main, if not only, witnesses in proceedings that may result in adolescents being sentenced to socio-educational measures.

Extra-legal punitive practices: the extermination of children and teenagers

Death squads, or "extermination groups," as they are called in Brazil, were created as early as the 1950s in Rio de Janeiro by civil policemen who were awarded bonuses for hunting and eliminating *outlaws*, i.e., robbers and thieves deemed beyond recovery. In the late 1960s, the mentality that poor black people were potential criminals and that outlaws could simply be eliminated had become widespread. Policemen and former policemen, retired members of the army, private security guards and even small shopkeepers started operating under names such as *esquadrão da morte* (death squad), *polícia mineira* (bloodsucker police), and *mão branca* (white hand), especially in the metropolitan region of Rio de Janeiro, which came to resemble a huge cemetery in the late 1970s because of the staggering number of corpses dumped there.[4] By the 1980s these groups had expanded under the protective wing of various retail businesses, and a number of important entities: senior police officers, court employees, public prosecutors, judges, politicians, and businessmen. According to representations that depict them in a favorable light, these groups were responsible for restoring order, cleaning up an area, and protecting the population. In practice, however, they acted as professionals who deal in death, seeking profits or non-material gains, such as notoriety and admiration in the eyes of the population (Misse, 1982, 1999; Alves, 1998; Souza, 1997).

The 1980s, when the activities of such squads reached their peak, have been dubbed the "lost decade" by economists because of the economic stagnation, inflation, unemployment, and sharp increase in inequality and poverty rates. The decade also saw great changes in trends in criminality: crime became more organized, violent crimes increased, and the overall sensation of a lack of security grew (Coelho, 1988; Adorno, 2002). Offenses committed by minors in the city of Rio de Janeiro included an increasing number of robberies. These outnumbered petty thefts, historically the most frequently recorded offense (Misse, 1999).

It was against this background of economic crisis and intensified violence that the extermination of young pickpockets or petty criminals, known as *pivetes*, became widespread.[5] There was also an increase in the number of minors who lived, worked, and practiced petty thefts on the streets, as well as an increase in the number of murders of these minors. The spread of these practices gained the tacit approval of segments of society frightened by the rise in increasingly violent crimes.[6]

---

4 The so-called "militias", groups of military police, former policemen and firemen who offer protection in small slum areas and low-income housing estates in return for monetary compensation, particularly in the western region of Rio de Janeiro, are therefore not recently created organizations.

5 Criminal social types are stereotypes that indicate social selection of criminal agents. Society, and not the police, created them (Misse, 1999). The term *pivete* was still current in the early twentieth century (Vianna, 1999).

6 An accusation made at that time by the National Movement for Homeless Boys and Girls (*Movimento Nacional de Meninos e Meninas de Rua* – MNMMR) in newspapers denounced the involvement of judges and court employees in the extermination of street children in Duque de Caxias, a municipality in the metropolitan region of Rio de Janeiro.

During the 1980s, therefore, the treatment of young offenders or alleged offenders from disadvantaged backgrounds deteriorated and reached a critical low as the killing or extermination of this section of the population won social approval. This attitude, which came to affect even younger children, was unprecedented in the history of the treatment of young offenders in Brazil. This type of extermination far exceeded the historical neglect of poor living conditions and social indifference in relation to all sorts of violence suffered by these populations. Of particular note is the physical violence perpetrated in juvenile detention centers and the arbitrary police approach in the form of torture and abuse, sometimes resulting in the killing of children and teenagers under the pretext of "taking justice into one's own hands" (self-help) and keeping future outlaws or *pivetes* off the streets. However, in addition to these representations found in the accounts of these deaths, it should be recalled that at that time the private security market was emerging as a very profitable business because of a rising demand for protection by shopkeepers. For the corrupt policemen involved, this was undoubtedly a much more worthwhile market than simply practicing extortion against *minors*, forcing them to share the profits of their thefts (Ferreira, 1979).

This reversal of roles, in which policemen violate legal and moral structures and become criminals themselves, was intensified during the 1980s as there was money to be made from drug-related graft and because this illicit behavior went largely unpunished. These events shaped the experiences and perceptions that poor children, teenagers and adults living in slums and housing projects had of police forces (Zaluar, 1985).

Hence, the greatest problem concerning the police was not their absence as an effective arm of the law in the management of local conflicts, but their very peculiar presence. This perverted approach is particularly apparent in interactions with juvenile delinquents; in combats that resemble urban warfare; in the retribution and retaliation by the police (which increased with extermination groups); and in the illegal trading in State resources for the police's own benefit (Misse, 1997). Therefore, not only was the police's legitimacy as a law enforcement agency undermined, but also youths and adolescents, especially those empowered by the drug trade based in the city's poorer areas, began perceiving them as an enemy who should be killed, or on occasions, as we shall see later, someone with whom they had to negotiate.

Under these circumstances, formal law and its enforcement as a means of dissuasion are unable to compete against private modes of conflict resolution founded on reciprocal violence. The latter will prevail in retail drug-market transactions, which operate under the permanent threat of retaliation (Misse, 1999) and the use of private justice by drug-gang bosses. With the increasing participation of youths in drug trafficking and the race by drug factions to acquire weapons (Dowdney, 2003), ever-greater cycles of retaliation are fostered by machismo and a hyper-masculine warrior ethos (Zaluar, 1994). The result is that, among these youths, resorting to force becomes the key principle governing social relations (Machado da Silva, 2004).

"The worms": the police as an enemy

Because of their violent conflicts with the police and members of rival gangs, for many youths, especially those from low-income communities, death is more than just a looming possibility; it is something palpable. Not surprisingly then, death is a recurring theme in the accounts of juveniles confined in detention centers as it is constantly present in their daily lives. Homicides committed by local criminals, enemies, policemen, and occasionally the youths themselves, eventually take a toll and diminish the value of life. In general these youths are aware that they are at risk of dying at any moment. They must also learn to survive in the face of frequent threats by rival groups or the police.

According to detainees who were interviewed, when they engage in shootings against policemen, they risk being "marked" or "tarred" (*pichados*). Changing address or leaving the "life of crime" are options considered by those who are marked for death. Another option, according to those who reported they knew they were going to die, is to take as much advantage of the "life of crime" as possible while it lasts, filling their pockets with money and indulging in drugs and women.

Osmar, an 18-year-old inmate, worked in the drug trade and was involved in a homicide that received extensive media coverage. He was one of those who feared being killed by policemen and therefore spoke of his wish to leave the drug trade:

> Entering is easy, getting out is difficult. They're calling me back [to the trade] but I'm already marked by the five-os.[7] I have faith in God, but I'm afraid. It's difficult to sleep. I'm afraid of being killed by the cops. The police back in my area sent me a warning.
>
> Osmar, 18 years old, in custody for theft

"Being marked" by policemen is a symbol of the "criminal subjection" process described by Misse (1999) whereby policemen treat youths as if they were bearers of "criminal" subjectivities and were naturally inclined to become criminals.[8] As a result, the police expect criminal behavior

---

7 Urban slang for police officers.

8 The notion of criminal subjection (Misse, 1999) compresses labeling, stigmatizing, and typifying processes into a single social identity specifically linked to the

from these "marked" individuals, making them suspects even before evidence of incriminating practices has been gathered. By considering themselves "tarred", these youths in turn bring the practice full circle and incorporate this accusation in their own self-representation. The account below by Eric, age 17, in his third detainment on grounds of drug trafficking, illustrates this "criminal subjection" process.

> If I give up this life, I'll have to go because of the police, I've been marked already. After my first arrest, I felt relaxed. One day I went to meet my girl at school, with my friend, and the five-os got us. They pointed their guns at me. They had a Blazer [name of vehicle used by police]. One guy said: 'Come on, Little Eric.'
>
> They took us behind the school and placed a bag over my head, inside the school grounds. He beat me up. The guy said he wanted my load of drugs, the piece (the gun), and money. I said: "I've got nothing to lose." So he began banging my head against the wall.
>
> Then the school bell rang. I was handcuffed, my head was bleeding. He continued pounding me in front of the school. My girl saw it and called my mother. My stepfather came and cursed at the policemen. He challenged the cops to a fight. When we left the school, the policemen made us shout in front of everybody that we would never sell marijuana again in the school grounds. But we wasn't [*sic*] even selling anything.

When they comment on the police, the youths interviewed express disgust. "Worms", "cops", or "enemies". "They should all die," pronounced one teenager. The police are no longer seen as representatives of the State and the law but as a "generalized other", following the same logic as the "criminal subjection" process, only in the reverse direction (Misse, 1999).[9]

Instead of turning themselves in, many of these youths shoot to kill when confronting the police. The reason for this is that, besides feeling an aversion to the "enemy", they fear being killed or beaten by the police if they surrender instead of being legally detained and taken to a police station without being beaten. This fear is justified, as the police usually follow a "shoot first, ask later" policy when entering favelas or chasing suspects in the street, always alleging that they were shot at first.

---

incrimination process and not as a particular case of deviation. The notion was proposed in order to understand these processes in a deeply unequal society, such as in Brazil. The stages of subjection are accusation, the automatic assumption that the agent has a tendency to practice criminal activities, self-representation by an agent or representations by their reference groups that justify or acknowledge the impossibility of justifying such practices.

9 According to the author, the entire police force seems to have been enveloped by a cloud of distrust and, under the same logic of the criminal subjection process, has become a stranger. Policemen are pictured as "enemies", cruel, arbitrary, corrupt, and cowards (Misse, 1999).

By the mid 1990s, the Rio de Janeiro Military Police's Special Operations Battalion (*Batalhão de Operações Especiais* – BOPE) had gained substantial power and assumed a leading role in confrontations with young criminals. The BOPE was conceived as a "war machine to hunt down outlaws" (Soares *et al.*, 2006). In their operations in slums, they employ an armored car called the "big skull" (*caveirão*) after the battalion's symbol – a skull – which is emblazoned on the car as a not-so-subtle emblem of war and death. The battalion's reputation is built upon its honesty and its avowed hatred of and implacable action against outlaws and drug dealers, which distinguishes it from the rest of the police. Its mission, as chanted in its war cry, is "to walk through the slum and leave corpses on the ground" (Soares *et al.*, 2006).

According to the statements of incarcerated juveniles, rancor, hatred and vengeance are feelings that permeate the memories of their experiences with the police, whether it be the Civil Police (PC) or Military Police (PM) forces.[10] Beatings, threats, torture, maltreatment, extortion, and especially deaths of family members or friends create a desire for vengeance and are recurring justifications given by youths to explain their anger at the police and their retributive fury, whether immediately, in face-to-face confrontations, or in hatching plans to gain advantages in confrontations by, for example, using the element of surprise. In the words of the teenagers interviewed:

> My father was a drug dealer, the slum boss's right-hand man. All the goods were overseen by him. He died in a gunfight. When I saw my father on the ground, dead, shot, I thought: 'Now I'm going to rebel. I'm the one who will get those cops'. I would stay on rooftops, armed heavily, like FAP (heavy automatic rifle) waiting for them to come so I could shoot them.
>
> Júlio, age 18, in custody for robbery

> I got busted under article 157 (robbery) when I stole a Pajero at the international airport. There was a chase and a gunfight. I was with three drug dealers over 18. [Interviewer asks: Were you afraid?] Not a bit. I have faith in my protector. Things went wrong for me that day. When one of the older men got out of the car, the cop shot him in the chest. I saw him fall by my side. I felt grief in my heart. It was really painful. I ran so as not to get shot too. I had a PT (pistol) and a charger with 30 bullets. I shot and hit the vehicle (…) Before that I had already seen a friend being shot at a place where drugs are sold ("boca").
>
> Luiz, age 16, in custody for robbery

---

10 In Brazil, the police are divided into two different forces, the Military Police (PM), which is in charge of overt and preventive patrolling, and the Civil Police (PC), also known as the judiciary police, in charge of investigating crimes.

These deaths spark a feeling of injustice and revolt and the desire to "honor the deceased" in the war against the "worms." Because of this hatred, "shooting at the cops" (*dar tiro em PM*) becomes an enjoyable practice, and managing to shoot a police vehicle or policeman is a badge of honor for many youths.

When describing the invasion of a *favela*, Jefferson, age 18, who was held in a detention facility after committing robbery, provides a good example of a situation in which a policeman was murdered in retaliation for the death of a fellow or "friend," a term employed to define anybody belonging to the same criminal faction. The gunfight with the police and the death of several "friends" were described very emphatically by Jefferson:

> Four of us were killed by the BOPE unit. We were moving out in a convoy of 80 people from H., M., and A. (names of slums).
>
> [Interviewer: How did it feel to invade another slum?!]
>
> It felt good. Shooting is awesome. It's like an adventure.
>
> I killed a BOPE policeman and I laughed. That guy had killed my friend; he shot him in the face. Another partner of mine was shot in the leg by the policeman. That guy was about to kill my friend, but I killed him first. I was behind him when he was about to shoot this boy. I saved my partner.
>
> We only left when the BOPE unit took over the hill. I was using a .30 caliber, the 'Pepê and Neném'. We exchanged fire for four days, holed up in houses. Five friends died.
>
> Jefferson, age 18, in custody for robbery

Like Jefferson, many other youths admitted to having committed violent acts against enemies and "worms" in retaliation for other acts of violence. In this vicious circle of armed attacks between rival groups of criminals and between criminals and the police, dignity and male honor are constantly tested, prompting these youths to react violently to the slightest provocation. When a youth is challenged or humiliated by rival drug dealers or the police, he feels compelled to use force and to defend himself according to the "hyper masculinity ethos", plunging himself into a cycle of armed warfare and violent retaliation. Defending and disputing territories thus becomes an obligation for those who take part in this logic of disputes over space and weapons, which are a symbol of this ethos (Zaluar, 1994: 109).

To these youths, the mentality of war prevails over all others. Machado da Silva (1996) aptly described it as atomized and segmented, instilling in subjects no objective other than remaining alive and seeking vengeance. As in the violent interactions with rival groups, a vicious circle of reciprocal violence is also established with the police (Zaluar, 1994).

Zico, aged 18, described his killing of a military policeman and the abuse he suffered by military and civil policemen following his arrest. His account, retold in a furious tone, depicts this logic of retribution in action: the "thirst for vengeance" and the desire to take "justice" into one's own hands shared by a number of teenagers.

> I was coming back from a party on hill Y on a motorcycle and going to slum Z. On the way, I killed a military policeman from the Xth Battalion. He had beaten me up once and killed a friend of mine. As I drove past him, I recognized this guy. So I went back just to shoot him. I killed the worm.
>
> But then I ran up against a patrol. I tried to escape but they caught me. These guys beat me. They stepped on me. They wanted to turn me in to the enemy from hill C [*controlled by a rival faction*]. They sent a radio message over there. They almost killed me. They said, "You're lucky. We didn't kill you because it's daylight now."
>
> They tortured me in the police station with electric shocks on my penis. They hit me a lot. I was a wreck. They even dunked me in water. These guys almost killed me. I guess it was because my family turned up there.
>
> [*Interviewer: Was the judge aware of this?*]
>
> No, I didn't tell the judge anything. I prefer to do it myself. I'll do it with my own hands. I'll fire right away.
>
> Zico, age 18, in custody for theft followed by homicide

It is clear that the justice system is perceived as failing to control and repress the police's criminal behavior. This perception is widespread within the lower-income population, who strongly resent this violent and unlawful action by the State. For some authors (Das & Poole, 2004), the State acts in margin areas as a "state of exception," defined as an essential complement to the sovereign state. This means that, at the margins of the State, uncertainty prevails regarding the enforcement of the law or, at best, there is a blurring of the distinction between legal and extra-legal punitive practices.

The inner workings of the State's operations at the fringes were exposed in the fictional narratives published in the book *Elite Squad* and in the enormously popular film of the same name.[11] A BOPE sergeant is killed by sniper fire in a confrontation with a drug gang. The whole unit is overcome by a feeling of revolt and a desire for revenge. A group of officers is in favor of immediate retaliation but is restrained by their commander, who fears the likely political repercussions of such a response. Revenge is nonetheless exacted by a group of the policeman's colleagues, who conduct an undercover foray to the murder site in unmarked cars. According to

[11] The film *Elite Squad* and the eponymous book depict the actions of the BOPE and Rio de Janeiro's Military Police in low-income areas from a police officer's point of view and contain an assessment of the Public Security system.

the report, "eight outlaws were executed so that justice was done." The motivation behind this act and the *modus operandi* were no different from the motivation or *modus operandi* of the acts of revenge carried out by groups of young criminals or by extermination groups that were responsible for the slaughter and indiscriminate murder of slum dwellers in the 1990s.

If the State, the provider of formal justice, is unable to protect slum dwellers, what then can be expected from it in the case of juveniles who are at "odds with the law"?[12] This group will be particularly vulnerable to actions by a state of exception. Thus, the idea of the State wielding power and applying rules within fixed boundaries is no longer valid as the scope of the State will vary according to the different ways of defining what and who is included in it (Asad, 2004).

Carlos, age 18 and in custody for robbery, knew that he could never be a victim in the eyes of law. He was very much aware of his own exclusion. Remarking on abuse by policemen, he said he did not believe they would ever be punished:

> I've already been tortured and beaten by the civil and military police. But if I denounced a policeman, do you think the judge would believe me? That's an outlaw's life: jail, beatings or death. Or else be put in a wheelchair.
>
> Carlos, age 18, in custody for robbery

Carlos described in his own words what has been termed "a bare life", a life so worthless it is treated as dispensable (Agamben, 2007), one that only stands a chance of being valued on the off chance that a corrupt policeman finds it profitable.

Bribes as political merchandise

Despite their aversion to the police, youths from poor areas allege that negotiating and paying bribes to military policemen has become a regular practice. The number of times youths were caught by the military police far outnumbered the times they were taken to a police station. Questionnaires given to the 105 inmates of the *Little House* detention center asked about the number of times they had been detained by the police. Before replying, inmates frequently asked, "Does this include the times when the police let me go?"

When asked about the number of detentions, one youth reported, "It depends. The cops got hold of me six times, but they released me because I gave them money. If you mean truly detained, as in taken to the police station, seven times in my case."

12 "At odds with the law" is the legal term for juvenile delinquents in accordance with the new approach proposed by the ECA.

Youths who work in the drug trade and those who are not closely connected to factions almost always talk their way to freedom with the agents of the law. Sometimes the policemen themselves will ask teenagers to steal specific items for their own use, such as digital cameras and cell phones. The outcome is that many policemen not only look the other way when teenagers commit crimes, but also take advantage of them.

Jonathan, age 15 and detained for robbery, experienced a number of situations in which he paid bribes to the police:

> You know, the policeman got hold of me, but he was corrupt and I already knew that. [This happened] down in one of Rio's southern districts. So I turned around to face him and you know what I said? "Listen man, we have to work this one out. I'll give you a cell phone and a 300-real small screen (digital camera), for you to let me go." So I gave them to him right away, I had the stuff.
>
> So I said: "So, can I go ahead and steal there? Am I safe? How about this? You catch me in the street and you let me go and keep my stuff." So he said: "Agreed, go ahead." There were two cops (PMs). I gave them everything, for them to share. Everything I had stolen. But he let me steal right there. So I told him: So, I'll steal there and it's cool, we're even?" And he said, "We're cool."
>
> So he let me go ahead. So I kept at it, right? He had already let me go…So I went to one of the side streets and began to steal. I started holding people up, taking things. Everyone would shout, "Police!" Then they'd come and let me go. Things had worked out. They were pretending that they were unable to catch me and when we met later…when we met he said: "Well then, it's ok, did you get what you wanted?" "I did." "Go then." And I ran back to the hillside (morro). [Interviewer: And you never saw the policeman again?] No, you know, I would always talk to him. He came and talked to me.
>
> Jonathan, age 15, in custody for robbery

Just as Jonathan concluded his story about paying bribes, another youth joined the conversation. "He [the policeman] is always in front of the X nightclub. These guys all take bribes," he said, meaning that he had also dealt with the same policeman.

If policemen are willing to negotiate with petty thieves for stolen goods and pay-offs, youths working in the high-turnover drug trade or who have the support of communities represent an even more attractive possibility. Bribes in these cases can be large enough to pay off an entire police unit, meaning that on the day that unit is on duty the *favela* need not fear surprise operations in their territory. This naturally implies that, according to youths, "clean" units are the ones that usually carry out the bulk of operations.

The corruption level in some police battalions is said to be exceedingly high. According to some accounts by youths, "bosses" have been known to pay monthly grafts to local battalion chiefs ranging from R$50,000 to R$250,000.[13] And this figure does not include revenue from the black-market sale of police weapons to drug dealers.

As a result of bribery, in some favelas the police reduce the number of drug raids and may call drug dealers in advance to tip them off about police operations planned by the battalion. It is no surprise then that police intelligence operations often fail. The following is an excerpt from an interview with three youths detained at *House*. Anderson (A), age 18, Luís (L), age 16, and João Vitor (JV), age 18, lived in different *favelas* but had at some point bribed policemen:

> A: The police warn us if they are going to enter the slum without having been given a bribe. We've been raised in the world of drug dealing. They don't come into my favela. They're afraid. That's mad! You can get caught much more easily on the streets (outside the slums).
>
> L: The dealers used to drop off more than 15 thousand [reais] at the police unit at the slum just to keep things nice and easy on weekends. Sometimes they come here to snort [cocaine], but they send word by radio and it's all OK. I've already seen a patrol [officer] in front of a funk music party selling some pieces (weapons). But sometimes the boss sends word to stop and says 'bribes are bullets.'
>
> JV: Their advantage is the armored car. But G3 bullets can leave marks on the "big skull" (armored car).

Hence, policemen make use of what Misse (2006) termed "political merchandise" when they receive money illegally from drug dealers and robbers. At this point, police corruption is no longer a mere deviation of conduct; it has evolved into a far-reaching illegal market that thrives on the drugs trade. This is a transaction in which power granted by the State to the policeman is appropriated for private benefit. Police corruption is only one type of "political merchandise"; others include activities such as blackmail and the sale of protection by armed groups, known as militias. A corrupt police force thus contributes to impunity and the maintenance of a network of illegal relations involving a range of criminal activities in Rio de Janeiro, including offenses by youths.

Legal punitive practices and the implementation of UPPs

Official police statistics (later contested by the public body that produced them) reveal that since 2005 there has been a decline in the number of children and teenagers detained by the police in Rio de Janeiro (see figure 1). This could reflect either an increase in extortion or a decrease

---

13 See US dollar exchange rate.

in the involvement of these youths in the drug trade (Silva, 2006), the latter being quite plausible as a sharp drop in retail drug activities in the city's poor areas has been observed since 2005 (Vargas & Misse 2010).

Figure 1. Individuals under 18 years of age arrested, killed or allegedly killed in confrontations with the police in Rio de Janeiro between 2000 and 2010

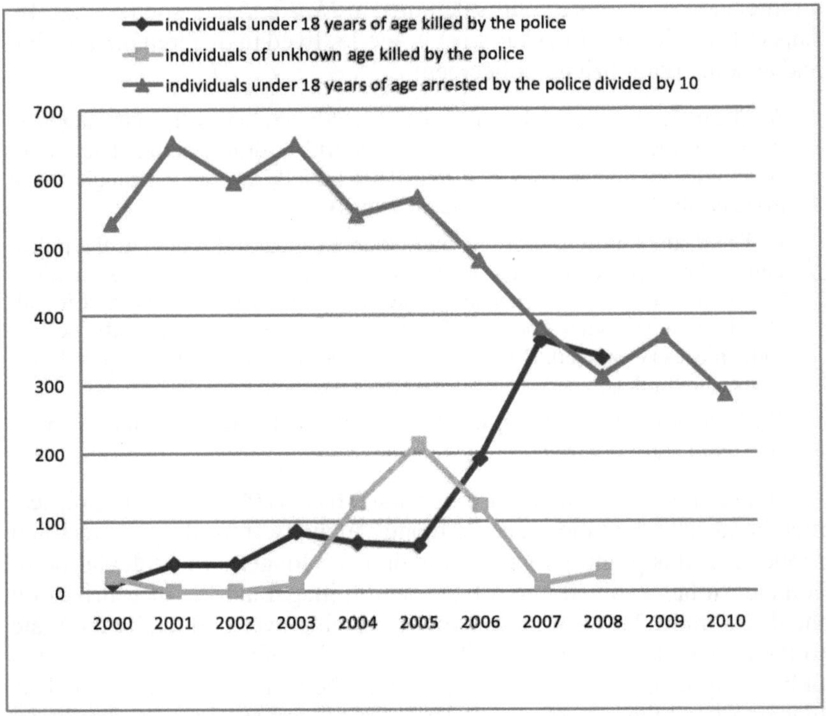

Source: Instituto de Segurança Pública – NECVU.

Nonetheless, data on police killings – the so-called *autos de resistência*, a category employed by the police to classify resistance or alleged resistance to police action that in practice can serve to cover up and justify homicide as legitimate defense – suggest a more likely hypothesis. Simply put, the drop in the number of adolescents detained is a result of their death in actual or alleged confrontations with the police.

The graph suggests the existence of an inverse correlation between the number of children and teenagers detained by the police and the number of deaths among these youths. In other words, as the number of detentions drops, the number of deaths increases. Also of note is the fact that the number of deaths for which no information about the victim's age was available and the number of deaths of individuals under 18 years of age follow opposite trends. When these two figures are added, the total corresponds to 2,078 lives lost between 2002 and 2009.

Some of these deaths, as we have seen, are the result of reflex reactions in such confrontations, but the majority is a consequence of the desire for retaliation among those involved. As a number of studies seem to suggest, there is yet another cause for some of these deaths: summary executions carried out by policemen. These follow certain patterns, as shown by necropsies: shots to the back and the head, and an excessive number of bullet wounds. Another statistic that may also be indicative of this type of summary execution is the small number of police killed in confrontations compared with civilians killed by the police (Cano, 2003; Misse *et al.*, 2013). The treatment these homicide cases receive in the Criminal Justice System also follows a discernible pattern: consensual and coordinated decisions by the actors in the system not to investigate killings of individuals deemed to be *drug dealers* or *outlaws* or make any formal accusations in such cases (Misse *et al.*, 2013).

More recently, violent practices of this nature have given way in some areas to a new model spearheaded by UPPs.[14] The overt purpose of the UPP model, which is still in its experimental stage, is to end armed confrontation, regain control of areas controlled by the drug trade and establish permanent community police detachments. A recent assessment of the effects of the UPPs suggests that these include the end of armed conflicts and *de facto* control over territory, as well as a dramatic drop in the number of deaths, in particular those arising from police intervention (Cano, 2012).

In this regard, implementation of UPPs represents progress compared with previous police practices and is a departure from an implicit policy based on violent confrontation. Nonetheless, it should be recalled that invasions and confrontations have not been altogether eliminated by this model, as the preparatory stage in this model consists of "tactical interventions", which are intended to clear the way for the UPPs to be set up. It is no coincidence that, as pointed out in the study referred to earlier (Cano, 2012), the sharp drop in the number of deaths is directly related to the decline in the number of police killings. Fruit of a new, post-confrontation approach, interventions in selected *favelas* prior to the establishment of UPPs have involved little armed confrontation because

---

14 The UPP model is a policing project for slums that has been under way since late 2008 in poor areas located in Rio's southern districts (the city's more affluent area), in the downtown area, and in slums close to the venues that will host the next Olympic Games.

of the departure from the *favelas* of the youths involved in the drug trade before the arrival of the occupation forces and because of the casualties suffered by these youths.[15]

Studies of police behavior in UPPs have underlined the similarity between the strategies adopted by the police to occupy slums and control the population in these and the strategy used by former drug dealers (Barbosa, 2012). The most striking aspect highlighted in these studies, however, is the as yet unresolved mistrust of the police among members of the community, particularly youths. While the general perception among youths is that life improved with the arrival of the UPPs, the question of violence is still a recurring theme in their comments voiced together with the demand for respect, especially from the police, whose presence and control of daily life is seen as more frequent and invasive (Abramovay-Castro, 2014).

Concluding remarks

Since 1990, Brazilian legislation covering juvenile delinquency has adopted a preventive/educational model in line with international conventions for the protection of children's and adolescents' rights. The change in legislation was a result of the return to democracy and the efforts of children's rights movements, which at the time were fighting against the extermination of young offenders.

We have sought in this article to show that the adoption of this legislation in the daily practices of those operating the criminal justice system, particularly the police, has been fraught with difficulty. The study focused on Rio de Janeiro, where police lethality is among the highest in the world. We have shown that the Rio de Janeiro city police record of violent, arbitrary behavior in slums (which very often leads to the summary killing of suspects), the daily lack of respect for the very poor and the frequent cases of criminal behavior have stimulated a sense of injustice and hatred, as well a desire for revenge, among slum dwellers, especially younger ones. Accounts of young slum dwellers and comments by the police show that they frequently get involved in cycles of retaliation and revenge. This police behavior, which is far removed from the law and even criminal, has contributed to youths having an extremely negative view of the police institution as a whole and reacting violently to it or even sometimes negotiating advantages. This type of police behavior is an important part of the process of social accumulation of violence in Rio de Janeiro (Misse, 1999).

---

15 All occupations have been publicly and widely announced by the government beforehand, which has led to the massive migration of people involved in the drug trade to other areas of the city or even to other cities in the state of Rio de Janeiro.

Since the main concern of this study has been to understand the reasons for the perpetuation of violent practices and the extermination of young offenders in Rio de Janeiro, it is important to examine the relationship between the practices of extermination groups in the past and the action nowadays in the fight against crime of the police, who, it should be noted, often shoot to kill and not infrequently execute young people summarily in the course of confrontations and incursions into *favelas*.

Various actors still insist on justifying death and executions as social cleansing. In 2008, a high-ranking Military Police officer compared youths involved in the drug trade to mosquitoes (echoing the fear caused by the dengue-fever epidemic plaguing the city) and considered the police an excellent social insecticide capable of killing every single mosquito.[16]

However, unlike previously, when extermination implied a shameful cleansing aimed at a social type that could still be deemed a possible victim of society, the war against the drug trade and drug dealers has cast aside any such concession. Currently, elimination is considered justifiable by the perpetrators and accepted as legitimate by the population, in a Manichaean clash between "good" and "evil" that the media has played its fair share in fomenting and disseminating.

The increasing use in Brazil of the fear of crime for political and media purposes and attempts to reform existing legislation in line with international reforms reflect the changing attitude to security and the abandoning of the model for the protection of vulnerable youths, which was never fully implemented in Brazil. It is against this background that the criminalization, punishment and even elimination of young offenders have gained strength.

Hence, poor black children and teenagers who are seen as potential *outlaws/* "*pivetes*" whether or not they have committed any crimes end up becoming *outlaws/drug dealers* regardless of their age. Another difference compared with the earlier period is that this generation of youths, empowered by resources made available by the illegal cocaine and weapons market, fights violently and to the death in a full-blown conflict against the enemy, whether it be a rival gang or the police, as in their social memory surrendering usually means death or suffering.

The police, on the other hand, whether working "to fight crime" or as part of the new program for slums known as UPPs, still exert a strong influence on the lives of people living in these slums, particularly youths, young offenders and alleged offenders.

Studies of this program have shown that it does not represent an effective break with the previous model. The fact is that police occupation has been

16 Article published online on the *O Globo* newspaper website, April 16, 2008.

the main mode of intervention in poor areas, and little importance has been attached to guaranteeing the rights of residents and youths in these areas.

References

Abramovay, M. Castro, M. (2014). "Juventudes, Violências e o Estado: Jovens em territórios com o programa 'Unidades de Polícia Pacificadora' no Rio de Janeiro". Número Especial – *Juventude, Violência e Controle Socioespacial na França e no Brasil*. Dilemas Revista de Estudos de Conflito e Controle Social, Vol. 5.

Adorno, S. (1999). "O adolescente e as mudanças na criminalidade urbana". *São Paulo em Perspectiva*, 13(4), 62-74.

Adorno, S. (2002 jul/dec). "Exclusão socioeconômica e violência urbana". *Sociologias* 4 (8), 84-135.

Adorno, S. (2005). "Le monopole étatique de la violence: le Brésil face à l'héritage occidental". *Cultures & Conflits*, 59, 149-174.

Agambem, G. (2007). *Homo Sacer: O poder soberano e a vida nua*. Belo Horizonte: Editora UFMG.

Almeida, S. S., Silveira, M. L. S., Olímpio, V. L. G. (2003). *Violência Estatal e Experimentos de Resistência na Vida Social Brasileira*. Rio de Janeiro: FAPERJ.

Altoé, S. (1990). *Infâncias Perdidas: o cotidiano nos internatos – prisão*. Rio de Janeiro: Editora Xenon.

Alves, J. C. S. (1998). *Baixada Fluminense: A violência na construção do poder*. PhD thesis, USP, São Paulo.

Alvim, M. R. (1995). "Infância das classes populares: A constituição da infância como problema social no Brasil". In: Abreu, A.; Pessanha, E. (eds.). *O Trabalhador Carioca – Estudos sobre trabalhadores urbanos* no Estado do *Rio de Janeiro*. Rio de Janeiro: JC Editora.

Alvim, M. R. (2001). "Meninos de Rua e Criminalidade: usos e abusos de uma categoria". In: Esterci, N. *et al.* (eds.). *Fazendo Antropologia no Brasil*. Rio de Janeiro: DP&A Editora, 17-340.

Alvim, M. R. and Valladares, L.P. (1988). "Infância e Adolescência no Brasil: uma análise da literatura". *BIB – Boletim Informativo e Bibliográfico de Ciências Sociais*, 26, 03-37.

Alvim, M.R. *et al.* (1992). *Da violência contra o "menor" ao extermínio de crianças e adolescentes*. Rio de Janeiro: NEPI/CBIA.

Anuário Brasileiro de Segurança Pública. Fórum Brasileiro de Segurança Pública, Ano 7, 2013.

Asad T. (2004). "Where are the margins of the state?" In: Das, V. and D. Poole, (eds.). *Anthropology in the Margins of the State*. Santa Fe and Oxford: School of American Research Press and James Currey.

Assis, S. G. (1999). *Traçando caminhos em uma sociedade violenta: A vida dos jovens infratores e seus irmãos não infratores*. Rio de Janeiro: Editora Fiocruz.

Batista, V. M. (1998). *Difíceis ganhos fáceis: drogas e juventude pobre no Rio de Janeiro. Instituto Carioca de Criminologia*. Rio de Janeiro: Freitas Bastos.

Barbosa, A. R. (2012). "Considerações introdutórias sobre territorialidade e mercado na conformação das Unidades de Polícia Pacificadora". São Paulo, *Revista Brasileira de Segurança Pública*, Vol. 6, No. 2, 2012, pp. 256-265.

Benevides, M. V. and Ferreira, R. M. F. (1983). "Respostas populares e violência urbana: o caso de linchamento no Brasil (1979-1982)". In: Pinheiro, Paulo Sérgio (ed.). *Crime, violência e poder*. São Paulo: Brasiliense.

Bugnon, G. and Duprez, D. (2010, jan/feb/mar). "Olhares Cruzados sobre o atendimento institucional aos adolescentes infratores". *Dilemas – Revista de Estudos de Conflito e Controle Social*, Rio de Janeiro, 3, 143-179.

Cano, I. (2003). "Violência Policial. Execuções Sumárias no Brasil – O uso da força pelos agentes do Estado". In: *Relatório de Execuções Sumárias no Brasil 1997-2003* – Justiça Global, Fundação Ford, 11-68.

Cano, I. (2012). *Os Donos do Morro: Uma Avaliação Exploratória do Impacto da Unidades de Polícia Pacificadora (UPPs) no Rio de Janeiro*. Fórum Brasileiro de Segurança Pública – LAV/UERJ. <http://riorealblog.files.wordpress.com/2012/07/relatc3b3riofinalcaf13.pdf>. Accessed 10 september 2012.

Chalhoub, S. (1990). *Visões da liberdade: uma história das últimas décadas da escravidão na Corte*. São Paulo: Companhia das Letras.

Cicourel, A. (2005). *The Social Organization of Juvenile Justice*. New Brunswick: Transaction Publishers.

Coelho, E. C. (1986). "A administração da Justiça Criminal no Rio de Janeiro: 1942-1967". *Dados – Revista de Ciências Sociais*, Rio de Janeiro, Iuperj, 29(1), 61-81.

Coelho, E. C. (1988). "A criminalidade urbana violenta". *Dados – Revista de Ciências Sociais,* Rio de Janeiro, Iuperj, 31(2) 145-183.

Das, V. (2008). "Los significados de seguridad en el contexto de la vida cotidiana". In: Das V. *Sujetos del dolor, agentes de dignidad.* Bogotá: Universidad Nacional de Colombia.

Das, V. and Poole, D. (2004). *Anthropology in the Margins of the State*. Sante Fe and Oxford: School of American Research Press and James Currey.

Dimenstein, G. A. (1990). *Guerra dos Meninos: assassinatos de menores do Brasil*. São Paulo: Editora Brasiliense.

Dowdney, Luke (2003). *Crianças do tráfico: um estudo de crianças em violência armada organizada no Rio de Janeiro*. Rio de Janeiro: Sete Letras.

Estatuto da Criança e do Adolescente, Lei No. 8.069/1990. <http://www.planalto.gov.br/ccivil_03/leis/L8069.htm>. Accessed 02 june 2012.

Ferreira, R. M. F. (1979). *Meninos de Rua: Valores e Expectativa de Menores Marginalizados em São Paulo*. São Paulo: Comissão de Justiça e Paz CEDEC.

Foucault, M. (1977). *Vigiar e Punir*. Rio de Janeiro: Vozes.

Gregori, M. F. and Silva, C. A. (2000). *Meninos de Rua e Instituições: Tramas Disputas e Desmanche*. São Paulo: Contexto.

IBASE/MNMMR/NEV-USP (1991). *Vidas em Risco: Assassinatos de Crianças e Adolescentes no Brasil*. Rio de Janeiro: IBASE.

Leite, M. P. "Da 'metáfora da guerra' ao projeto de 'pacificação': Favelas e políticas de segurança pública no Rio de Janeiro". *Revista brasileira de segurança pública*. São Paulo, Vol. 6, No. 2 set 2012, pp. 374-389.

Lima, R. K. (1989). "Cultura jurídica e práticas policiais: A tradição inquisitorial". *Revista Brasileira de Ciências Sociais* 4 (10), 65-84.

Machado Da Silva, L.A. (1999, november). "Criminalidade Violenta: Por uma nova perspectiva de análise". *Revista de Sociologia e Política*, 13, 115-124.

Machado Da Silva, L. A. (2004, june). "Sociabilidade violenta: por uma interpretação da criminalidade contemporânea no Brasil urbano". *Sociedade e Estado* 19(1), 53-84.

Machado Da Silva, L. A. (2008). "Violência urbana, sociabilidade violenta e agenda pública". In: Machado Da Silva, L. A. (ed.). *Vida sob cerco: violência e rotina nas favelas do Rio de Janeiro*. Rio de Janeiro: Nova Fronteira.

Marinho, F.C. (2012). *Mudanças, Resistências e Mediações no Campo Sociojudiciario da Gestão da Delinquência Juvenil: Uma abordagem comparativa entre a França e o Brasil*. PhD thesis, UFMG, Belo Horizonte.

Misse, D. G. (2013). *Políticas sociais em territórios pacificados*. PhD thesis, PPGSD, UFF, Niterói.

Misse, M. (1982). "Direito e Conflito Social em Nova Iguaçu". In: Miranda Rosa, F.A. (coord.). *Direito e Conflito Social no Brasil*. Rio de Janeiro: FINEP – CEJUR.

Misse, M. (1997). "As ligações perigosas: mercado informal ilegal, narcotráfico e violência no Rio". *Contemporaneidade e Educação* 1(2), 93-116.

Misse, M. (1999). *Malandros, Marginais e Vagabundos: A Acumulação Social da Violência no Rio de Janeiro*. PhD thesis, IUPERJ, Rio de Janeiro.

Misse, M. (2006). *Crime e Violência no Brasil Contemporâneo. Estudos de Sociologia do crime e da violência urbana*. Rio de Janeiro: Lumens Juris.

Misse, M. (2007). "Notas sobre a sujeição criminal de crianças e adolescentes". In: João Trajano Sento Sé &Vanilda Paiva (eds.). *Juventude em conflito com a lei*. Rio de Janeiro: Garamond.

Misse, M. (2010). "Crime, sujeito e sujeição criminal: Aspectos de uma contribuição analítica sobre a categoria 'bandido'". *Lua Nova*. São Paulo, 79 15-38.

Misse, M.; Grillo, C. C.; Teixeira, C.P.; Néri N. A. (2013). *Quando a polícia mata: Homicídios por autos de resistência no Rio de Janeiro. (2001-2011)*. Rio de Janeiro: Booklink.

Misse, M. and Vargas, J. D. (2010). "Drug use and trafficking in Rio de Janeiro: some remarks on harm reduction policies". *Vibrant* (Florianópolis), Vol. 7, 88-108.

Misse, M. and Vargas, J. D. (mar/apr 2009). "A produção decisória do sistema de justiça criminal no Rio de Janeiro ontem e hoje: um estudo preliminar". *Revista Brasileira de Ciências Criminais*, 17 (77), 237-260.

Misse, M. *et al.* (1973). *Delinquência Juvenil na Guanabara: uma introdução sociológica*. Rio de Janeiro: Tribunal de Justiça da Guanabara e Juizado de Menores da Guanabara.

Neri, N. E. (2009). *Tirando a cadeia dimenor*. (Master's thesis) PPGSA/IFCS/UFRJ. Rio de Janeiro.

Neri, N.E. (2012). "'Ninguém é peito de aço' um olhar sobre as relações entre jovens em conflito com a lei e policiais no Rio de Janeiro". In: Misse, M.; Werneck, A. (ed.). *Conflitos de (grande) interesse. Estudos sobre crimes violências e outras disputas conflituosas*. Rio de Janeiro: Garamond, 205-225.

Passetti, E. (1996). "Crianças carentes e políticas públicas". In: Priore, M.D. *História das Crianças no Brasil*. São Paulo: Contexto, 347-375.

Pinheiro, P. S. (ed.) (1999). *Continuidade autoritária e construção da democracia*. FAPESP/CNPQ/FORD. <http://www.nevusp.org/downloads/down000.pdf>. Accessed 1 april 2012.

Rizzini, I. (2000). *A Criança e a Lei no Brasil – Revisitando a História (1822-2000)*. Brasília, DF: UNICEF; Rio de Janeiro: USU Ed. Universitária.

Rizzini, I. (1995). "A arte de governar crianças". In: *A arte de governar crianças: a história das políticas sociais da legislação e da assistência no Brasil*. Rio de Janeiro: Instituto Interamericano Del Niño, Editora Universitária Santa Úrsula, Amais Livraria e Editora, 99-166.

Rizzini, I. (1991). *Infância e pobreza no Brasil: Uma abordagem sobre a contribuição acadêmica na década de 80*, brochure.

Sento-Sé, J. T. (ed.) (2003). *Perfil dos jovens em conflito com a lei no Rio de Janeiro*. Rio de Janeiro: CESeC/UCAM e LAV/UERJ.

Sinhoretto, J. (2001). *Os justiçadores e sua Justiça. Linchamentos, costume e conflito*. Dissertação de Mestrado. FFLCH, USP.

Silva, E. R. A. and Gueresi, S. (2003). *Adolescentes em conflito com a lei: situação do atendimento no Brasil*, texto para discussão No. 979, Brasília: IPEA.

Silva, J. S. (2006). *Caminhada de crianças, adolescentes e jovens na rede do tráfico de drogas no varejo do Rio de Janeiro, 2004-2006*. Rio de Janeiro: Observatório de favelas. <http://www.observatoriodefavelas.org.br/userfiles/file/Pesquisa%20 Trajet%C3%B3ria%20de%20Crian%C3%A7as,%20Adolescentes%20e%20

Jovens%20na%20rede%20social%20do%20tr%C3%A1fico%20de%20drogas%20no%20rio%20de%20janeiro.pdf>. Accessed 5 april 2012.

Silva, G. M. (2010). *Ato Infracional: Fluxo do Sistema de Justiça Juvenil em Belo Horizonte*. (Master'sthesis). Programa de Pós-Graduação em Sociologia FAFICH – UFMG.

Soares, L. E., Batista, A., Pimentel, R. (2006). *A elite da tropa*. Rio de Janeiro: Objetiva.

Sousa, J. A. (1997). *Os Grupos de Extermínio em Duque de Caxias – Baixada Fluminense*. (Master's thesis). PPGSA/IFCS, Rio de Janeiro.

Sposato, K. B. (2006). *O Direito Penal Juvenil*. São Paulo: RT.

Vargas, J. D. and Marinho, F. C. (jul/dec 2008). "O Programa Liberdade Assistida em Belo Horizonte". *Educação & Realidade*, 33 (2), 147-162.

Vargas, J.D. and Misse, M. (2008). "L'évolution de la consommation et du trafic de Drogues illicites à Rio de Janeiro". *Déviance et Societé*, 32 (3), 377-391.

Vargas, J.D. (2011). "Práticas punitivas legais e extra-legais relativas à infância pobre e delinquente no Rio de Janeiro". In Barreira, C.; Sá L.; Aquino, J.P. (ed.). *Violência e Dilemas Civilizatórios. As práticas de punição e extermínio*. Campinas: Pontes, 179-202.

Vianna, A. R. (1999). *O Mal que se Adivinha: Polícia e Menoridade no Rio de Janeiro, 1910-1920*. Rio de Janeiro: Arquivo Nacional.

Violante, M. L. V. (1982). *O dilema do decente malandro*. São Paulo: Cortez.

Zaluar, A. (1985). *A máquina e a Revolta. As organizações populares e o significado da pobreza*. São Paulo Brasiliense.

Zaluar, A. (1994). *Condomínio do Diabo*. Rio de Janeiro: Editora da UFRJ.

Zaluar, A. (1998). "Para não dizer que não falei de samba: os enigmas da violência no Brasil". In: Schwarcz, L. M. (ed.). *História da vida privada no Brasil*. São Paulo, Companhia das Letras, Vol. 4, pp. 245-318.

Zaluar, A. (2004). *Integração perversa: Pobreza e tráfico de drogas*. Rio de Janeiro: Editora da FGV.

Zaluar, A. (2007). "Democratização Inacabada: fracasso da segurança pública". *Estudos Avançados* 21 (61), 31-49.

Zilli, L. F. and Vargas, J. D. (2013). "O trabalho da polícia investigativa face aos homicídios de jovens em Belo Horizonte". *Ciência e Saúde Coletiva* (Impresso) Vol. 18, pp. 621-632.

Part 3

Violence, Identities and Development

Regional Mining Identities vs. Multinational Mining Interests

The End of Traditional Small-scale Gold Mining in Marmato (Caldas, Colombia)[1]

Robert V.H. Dover, Paula A. Hinestroza Blandón & Gloria Patricia Lopera Mesa

Development, participation, and identity

We have described the predominant development paradigm elsewhere (Dover, *et al.*, mss.) as a phenomenon that convokes specific and inequitable social relations that center themselves around the social, political and environmental circumstances and conditions that major development projects – infrastructure and resource extraction – generate. At the same time, the paradigm also characterizes and evaluates in more general terms the different state, institutional, industry, and community actors based on how they position themselves relative to the paradigm, the degree to which they are able to, allowed to or want to participate in it, and the legitimacy of their discourse/epistemology regarding development – greater social distance meaning less dominant society social recognition and/or value.

Following Escobar (2010) and Ramírez (2010), we recognize that there is a normative and prescriptive aspect to this paradigm, an idealization of the participating citizen and their social context that follows or is coherent with the predominant map of social relations (class, ethnicity, gender, etc.), and an expectation of conformity to the ideology of progress that

1 This article is a result of the research project, "La consulta previa como herramienta para la transformación de contextos de desarrollo y la mitigación de conflictos socioambientales" financed by Colciencias (Colciencias 111552128789), the Universidad de Antioquia – CODI, and the Universidad EAFIT, and developed by the research groups: RERDSA (INER, Universidad de Antioquia); and Justice & Conflict (Universidad EAFIT), and from the project, "Explotación minera en Marmato (Caldas): conflictos, reivindicaciones y movilizaciones en relación al desarrollo futuro de la localidad", financed by the Universidad de Antioquia – CODI.

We would also like to extend our appreciation to the Universidad de Antioquia, Medellín, Colombia, and their program, "la Estrategia para la Sostenibilidad de los Grupos de Investigación 2013-2014", for their support in the publication of this article.

the paradigm represents. Conformity to the paradigm map becomes one measure that the state and industry use to delimit the legibility of constituent populations, recognizable "others" that fit or are placed into development's normative narratives (Scott, 1998). Identity and sociocultural diversity become factors to domesticate through the mechanism of differentiated participation. The normative narratives create and distinguish different degrees of *otherness*, such that the *distant other* is mutually constructed and externally represented in a field of social and legal alterity as in the case of recognized ethnic groups, while the *near other* is internally constructed as an "autochthonous sociocultural, economic or political alternative" (Dover, *et al.*, mss.) but externally represented as part of the larger national identity narrative.

The paradigm also de-territorializes identities, not in the active actor-centered sense described by Gupta and Ferguson (1992) in their discussion about the disassociation of identity and territory, but rather as a consequence of a globalization process that obviates local identity construction and differences both in physical terms through forced or guided relocation, and/or the physical alteration of the social, cosmological and physical landscape, and in epistemic terms through the resignification of territory, the designation of the appropriate use of territory, and the assignation of permitted identities, a process Hale refers to as "neoliberal multiculturalism" (2005).

To the extent that the state as an advocate of the development paradigm resignifies territories and their uses, and assigns permitted identities instead of recognizing traditional or emergent ones, is the degree to which there exists latent and potential social, environmental and political impact that can create or exacerbate contexts of structural inequity, contribute to epistemic violence, and signal the possible subsequent incursion of or emergence of violent actors who take advantage of the circumstances and attach political meaning and intentionality to the inequitable context. The resignification and assignation of identities are self-evident in those cases where the *otherness* is ethnically based. But what happens when the *other* is not so distant from the national identity, where the basis for non-conformity lies less on sociocultural axes distinct from the dominant axis, and more on the construction of local autonomy and a local interpretation of development?

This article proposes to examine how development projects reconfigure and deterritorialize local identities of impacted populations through the normativization of inclusion through recognizable identities, and how, in the specific example of the dispute between small and large-scale gold mining in Marmato, Caldas, the development paradigm also resignifies territory and its appropriate use.

The case of Marmato was one of the studies in an 18-month project we conducted, 2011-2014, regarding the incidence and efficacy of Prior Consultation (PC) and other mechanisms of participation in vulnerable ethnic and non-ethnic populations in three contexts of infrastructure development and large scale resource extraction: first, the construction of a tunnel that connects Medellín with the coast of Urabá that affected a community of rural agriculturalists within the rural jurisdiction of the city of Medellín; second, the construction of a dam over the Rio Rancheria in the department of the Guajira that affected directly and indirectly indigenous and rural agriculturalist communities, as well as medium-scale rice producers; and the last – the focus of this paper – the granting of a gold-mining concession to a multinational corporation in the municipality of Marmato, an area that has a long tradition of small-scale and artisanal mining. One of the central issues to the project concerns the construction of identities, the mobilization of "correct" identities in terms of being able to access the PC process, and the politics of inclusion where inclusion is often at the cost of local and autonomous constructions of identity and the social and territorial definition of local "vital spaces".

In the particular case of Marmato, the analysis concerned the identification of socio-environmental conflicts between the local small-scale traditional mining community and the large-scale national and multinational mining companies[2] since 2007 when the Compañia Minera de Caldas first expressed interest in establishing an open pit gold mine in what has been the traditional mining section of the municipality.

We defined several lines of analysis each one with a different methodological base: a) review and analysis of the jurisprudence and norms regarding citizen participation and the regulation of mining projects; b) the identification of the actors involved and their positions regarding the tension between formal and informal mining and the specific circumstances of Marmato; and c) a description and analysis of the different scenarios and processes of negotiation, mobilization and/or articulation between the different actors.

What most marks this research is the transdisciplinary focus where co-researchers from anthropology and law were in constant dialogue. The final objective was to comprehend the sociocultural matrix of the Marmato inhabitants and the importance of their informal mining practice in this

[2] The previous Colombian Código Minero (Mining Code) of 1998 distinguished between the different levels of mining including small-scale mining as a legitimate level of mining subject to different regulations and management. The current Código Minero of 2001 does not distinguish the different levels of mining thereby taking away the specific legitimacy of small-scale mining, and gives it the designation of informal mining, a category that facilitates the dismantling of small-scale mining communities and traditions.

matrix; at the same time, recognize the legal framework within which the local socio-environmental conflicts developed, and the transformation of this framework as a result of the manner in which the law was used, reinterpreted, or ignored by actors with the political and legal authority to resolve these issues.

We initially chose the conflict between small and large-scale mining in Marmato, Caldas, because of the co-existence of an ethnic population (indigenous and Afro-Colombian) with a large mestizo population, but which subordinated these identities to one formed in their principal local economic activity – small-scale mining. In the course of the fieldwork, we were able to evidence, in contrast to the construction of inclusive local identities, how the presence of the multinational companies with the proposal to convert the region into a large-scale mining concern reconfigured and deterritorialized these identities, and how their use of epistemic violence and law changed the central concerns in local discourse and the social relations of local actors.

The article will first look at how the Colombian state has constructed the *other* and the differentiated participation rights of different *others*, and a recent Constitutional Court decision regarding citizen participation of a non-ethnic community in a development project that brings the central tenets of this discussion into question. We then describe the situation in Marmato and the conceptual parallelisms between the traditional socioeconomic activity of the local population and the community that was named in the Constitutional Court decision. The article ends with a discussion of the impact of the epistemic violence that results from the imposition of state constructions of the *other* on local communities.

The near and distant "other"

Colombia is a country where there are few legal and political mechanisms to stipulate an official identity beyond what is considered a national one. Those persons or communities that cannot do so are denied certain processes that are clearly identified as the recognition of special rights to compensate social and historical injustices on one hand, and to protect the integrity of their social being on the *other*. This has clearly been the case regarding the process of Prior Consultation (PC) with affected communities in the context of large-scale development, resource extraction, and legislation projects, and the exclusive use of this process on the part of the historically disadvantaged ethnic groups when it is used.

PC is the most complex and complete of a series of non-adversarial participatory mechanisms available to Colombia's vulnerable populations in the sense that, more than just a non-binding consultation, PC is intended to give ethnic populations a certain degree of leverage to negotiate the mitigation of impact, and incidence in the design and implementation of development and legislative projects that potentially affect them and their territorial and identity rights. PC first appears as an option to ethnic communities in the Political Constitution of 1991, and through Law 21 of 1991, which ratified the ILO Convention 169. In the 20+ years since its ratification, the degree of incidence that the ethnic populations enjoy has been subject to different political and juridical interpretations – from an unregulated right with no consistent or implemented procedure to an advisory consultation regarding impacts to the idea of informed consent with the power to veto. The current view taken by the Constitutional Court (CC) is that of *consentimiento libre, informado y previo* ("opportune, informed, and unencumbered consent"),[3] which, unfortunately, has not had the far-reaching impacts that were expected with the liberal interpretation of *consent*. In spite of the recent favorable CC rulings regarding the conditions of negotiation and the rights of the affected communities to be represented in the consultation process, PC as a mechanism of participation has been problematic for a number of reasons: the lack of political will on the part of the state to ensure compliance with PC; the inconsistent application of PC in different development contexts and based in part on local state and multinational actors, the interpreters of the process, and the degree of preparedness of the affected communities; the lack of consensus regarding the moment when the process should begin – during the exploration phase or when the bulldozers have begun to move earth –; the unequal playing field or differentiated access to information on the part of the multinationals and the state, and on the part of the affected communities; the language of negotiation that favors the development paradigm; the community recourse to "*vías de hecho*" or social protest and civil disobedience when legal resources do not work; and the fact that other vulnerable populations, the *near other*, that cannot claim an ethnic identity but which are equally affected are not included in the PC process (see Lopera and Dover, 2013).

The exclusivity of ethnic identity markers to indicate eligible groups and the public policy attempts to mitigate the impact of development on groups that claim an ethnic identity recently underwent an important reinterpretation through the CC's Decision T-348/12, on May 15, 2012.[4]

3 This perspective began with the CC decision T-769/2009 (the case of Mandé Norte) and continued with subsequent cases in which the court confirmed that in the event of high impact projects that threaten the survival of ethnic groups, prior consultation is not a sufficient guarantee, rather it is necessary to reach ("opportune, informed, and unencumbered consent"). This view of the CC has created a great deal of controversy in the government with recent pronouncements that this liberal interpretation of the CP process is paralyzing development.

4 Writ for the protection of constitutional rights ("acción de tutela") brought by the Asociación de Pescadores de las Playas de Comfenalco – ASOPESCOMFE, against the Distrito Turístico de Cartagena, the Consorcio Vía al Mar, the Ministerio de

The case cited concerns an association of coastal fishermen who fish on the beaches of Cartagena, and whose social cohesion and identity-base is centered on a shared economic activity and not ethnicity or kinship, confronting the construction of a highway which would adversely affect what is their principal economic activity. Citing from an April 20, 2012, technical concept offered by the Instituto Colombiano de Antropología e Historia (ICANH), the CC writes:

> Traditional fishing communities are populations that should especially be listened to regarding infrastructure projects that intervene in that space where they practice their trade. This because it concerns groups of people that are permanently dedicated to fishing to guarantee their food security and the economic sustenance of their families, and, in this sense, *the area of ocean or beach that they use to fish becomes a vital space.*
>
> …
>
> The only form in which the measures [to mitigate impact] are efficient and adequate, is that they be the result of an evaluation, not just of the negative impacts on the environment, but as well of the impact over the community, taking into consideration its special characteristics and its relationship with the environment, that is to say, it is necessary to realize a "native evaluation of impact" (CC, 2009).

What is significant in the CC's decision is the fact that the text they cite from the ICANH describes an authentic (in the sense of emergent and traditional) economic activity that is part of their "vital space" and contributes to their well-being, represented in the idea of food sovereignty, and in the specificity and local grounding of their socioenvironmental relationships, both which serve as identity markers. More importantly, the court recognizes the distinct epistemological base, a boundedness to their socioeconomic and historical experience, out of which they interpret differently, and in terms of identity, the impacts of development.

In effect, the CC states that *otherness* is not defined exclusively in terms of ethnicity, but rather to the boundaries that different social and historical experiences create which may or may not include ethnicity, what we have referred to as the "near" and "distant" *other* respectively (Lopera and Dover, 2013; Dover, *et al.* mss.):

> As a result, the CC will order the Consortium Via al Mar, taking into account the current impacts of the project, to guarantee spaces for negotiation with the [Fishermen's] Association, negotiations that take into consideration their

---

Ambiente y Desarrollo Sostenible, the Instituto Nacional de Concesiones – INCO – today the Agencia Nacional de Infraestructura –, the Dirección General Marítima – DIMAR and the Instituto Nacional de Vías – INVÍAS. Presiding magistrate: Jorge Ignacio Pretelt Chaljub.

opinions, as well as to design conjointly the compensations, **which should be in accordance with the character of their work as traditional fishermen, and not necessarily with the economic nature of what they do.**

In addition, the CC establishes that sociocultural vulnerability is more than ethnicity, and that sociocultural and sociohistorical identities are formed in collectively defined and experienced "vital spaces", in which different norms operate to conserve their relationship with their social and territorial context. This idea is not something new to the social sciences, but it acquires additional significance in light of the tendency of the CC to make new, binding social policy.

We begin this discussion citing the CC decision regarding folk fishermen in Cartagena because in the case of the conflict between small and large-scale gold mining in Marmato, not only are the territorial and identity rights of the Marmato indigenous and Afro-Colombian populations at risk, but in addition – and this parallelism with the Cartagena example is the central argument being developed here – of the majority of the population including the ethnic populations who define their sociocultural and sociohistorical identity on the basis of the exercise of their traditional economic activity, small-scale gold mining, as will be explained below.

The Case of Marmato

Marmato, Caldas, is a tiny municipality with a multiethnic population of 8,848 inhabitants (55% Afro-Colombian and 17% indigenous). The center of the municipality is located in the upper part of a mountain known as Cerro el Burro, reachable by a rather torturous unpaved road that rises from the populated area in the lower part known as El Llano, where most of the Afro-Colombian population is located. El Llano is also the proposed site for the relocation of the municipality in case the multinational mining company Gran Colombia Gold takes the option of open-pit mining. Although there is mining throughout most of the mountain, the Marmateños who practice traditional mining do so in the upper part of the mountain, and the large concentration of over 400 small-scale mines is found in the Cerro el Burro, in the neighboring sector of Cien Pesos, in the rural area of Echandía, and to the north in the areas of Guayabales and El Salto.

Municipality of Marmato, Caldas (Google Earth, March 12, 2013).

Small-scale mining constitutes the axis around which the Marmato local economy, social relations, the construction of identity and its material and symbolic representations is based. The mining activity itself as practiced by Marmateños is a very *sui generis* way of perpetuating the Marmato identity through subsequent generations.

To understand the importance of mining in Marmato, not just as mode of production and economic sustenance but also as a structural element in social relations and cultural practices, implies a recognition of the particular characteristic of the municipality where the mine entrances, houses, population, public buildings and spaces are all contiguous on the same mountain. It is also necessary to understand how the state has tried to include Marmato in the national and regional development scheme through a series of laws and decrees that try to condition the local mining; legislation that considers Marmato as a place rich in gold resources, but that do not consider that it is first a historical population and disregard the sociocultural factors that this implies. The way of life that is evidenced in the historical experience of the Marmateños has been threatened by external agencies or promoted by the state and materialized with foreign capital.

The development of the municipal center of Marmato, which is also the economic, sociocultural and political center, represents the intimate experience with small-scale mining of the local population. Previously, the population center was located in the village of San Juan to the south of Marmato, now under the administration of Marmato, and the people slowly began to relocate close to their mines, constructing their camps near the mine entrances located in the Cerro el Burro; this situation gradually transformed into the migration of the miners and their families to what is now the urban center of Marmato, and the subsequent construction of the town and its transformation into the municipal center. In this regard, the work place, the residence and the social spaces were configured and continue to be configured in a unique sociospatial relationship.

View of municipal center of Marmato and Cerro el Burro upper zone from the vereda (passage) of San Juan (Photo courtesy of Paula Hinestroza 2012).

The proximity of mines and residence has been key in the reproduction of the system of traditional mining. Considered locally to be largely men's work, nevertheless, children are socialized early into the mining activity by participating in different ways; taking food to the family members who work in the mines or processing plants, simply visiting the installations, as helpers when they are older in minor tasks that do not involve risks but through which they receive some remuneration, or in activities such as panning for gold in the waste material. At the same time, the integration of children into the mining activities is a direct result of the father/son relationship; fathers give their sons the trailings, the leftover sand that is a byproduct of the grinding of the stone that contains the gold ore, so that they process it again and extract that last remaining grains of gold for their own use. In this way, children become familiarized, in everyday spaces, with the mining activity and the economic benefits this generates.

Although the mining is concentrated in the municipal center, it extends to the rural areas as a complementary activity that allows the rural populations to alternate between mining and agricultural as a domestic economic strategy. In one way or another, the entire local population of Marmato is involved in mining because of the productive and sociocultural activities that this generates. Production implies not only work in the mine, but also work in the processing plants ("molinos"), where women work as well in the supervision of the work, the purchase and transport of materials to work with and that ensure the security of the mines – explosives, timbers, railcars, work tools – and, finally, the profits are invested in the local and regional economy. The social aspect is constituted by the family and fictional kin networks that are activated by the mining; they form associations and small and medium-sized mining concerns that involve various family members. They also are evident in solidarity networks that are activated in the case of accidents or death in the mines, moments when the inhabitants collect monies and other support to help the affected miners and their families.

In this regard, the gold resources that are found in the Marmato area allow the involvement of most of the municipal population and many more from the region in a system of traditional production that historically has guaranteed their sustenance. At the same time, they have made possible the formation of a cultural identity shared among the local population.

Vertical territoriality as a condition of social order based on the co-existence of small and medium-scale gold mining

One of the central elements in the manner the Marmateños have constructed their territoriality and distributed the exploitation of gold is the separation of the upper Zone A and the lower Zone B, each dedicated to the small-scale and medium-scale mining respectively, as established initially in Law 66 in 1946, and reiterated in Decree 2223 in 1954. This territorial distribution was adopted as a formula to resolve the social and economic problems that have their origin in the middle of the 19th century due to the granting of the exclusive right to a single person to extract the gold from the Cerro El Burro and neighboring sectors, at the same time to arrive at an agreement regarding this social problem with an interest to improve the efficiency in the extraction of the gold. Such a distribution could be maintained throughout the 20th century as a result of a set of political and economic circumstances that favored, or at least not get in the way of the development of small-scale mining in this part of the country.

The vertical distribution of the Cerro El Burro and surrounding sectors between the upper and lower zones constitutes a central element in the appropriation and representation of this space as a territory and is present in many of the material and symbolic practices that define their territoriality. In addition to the division into two different scales, small and medium, of mining production, this territoriality is present in other elements that structure social and production relations among Marmateños:

The lower zone B which has as its principal reference the medium-scale mine of Mineros Nacionales identifies with the development of an industrial and technical mining, and represents a place that guarantees permanent employment for Marmateños and people from outside the municipality, and formal labor conditions with the advantages (health and welfare benefits, professional risk) and disadvantages (keeping a full-time work schedule which includes night and holiday shifts, being subject to labor rules, a stable wage but lower than what an independent miner earns) that Marmateños associate with labor formality.

At the same time, the upper Zone A represents a space where economic and labor relations are characterized by informality at the same time that there is less asymmetry due to the fact that small mine owners tend to offer work to local miners which gives the latter a certain amount of freedom to negotiate their work conditions, especially because of the option to work in an independent manner in panning ("bareque") or the "guacha".

"Barequeo" or panning for gold (photo courtesy of Paula Hinestroza 2012).

This particular weaving of social and productive relations is what distinguishes Marmato from other mining communities that develop around enclave economies, resulting in what the sociologist, Melina Lasso, following Bourdieu, has referred to as a particular "productive habitus" of the Marmato miners (Lasso Lozano, 2011: 170-191), and the anthropologist, Carlos Julio González designates as a Marmato "mode of production, based upon the practice of traditional mining (González Colonia, 2012).

This distribution model regarding space and the mineral resource between the upper zone where the gold is for the Marmateños and other small-scale miners that come from other parts to mine as long as they follow the social rules established by the Marmateños, and a lower zone where the state can contract external mining companies that also must respect the social rules to earn the right to stay in Marmato and benefit from its resources, is firmly engrained in the way the Marmateños signify their relationship with their environment (González Colonia, 2012). This is evident in the explanation offered by don Eulises Lemus regarding the reason behind the legislation that reserved the upper part of the Cerro El Burro to small-scale mining:

> This is the mountain, there is the Cauca [river]. Above there is a hydrothermal deposit that formed by the rising of magma. A plate comes up and produces a change in elevation. This is hot. When it reaches the surface it fractures. There are lots of fractures in the rock. Afterwards a molten liquid passes through the fractures and these are the veins. The molten liquid of less quality is what reaches the upper part. For this reason it is divided into two parts. One part is the lower zone that the state has maintained as a national reserve. This is the richest part. But the gold is not superficial rather deep inside [the earth]. For this reason, the state turned it over to a multinational with the ability [to mine it]. The lower part is the more rich but it is deep. As it rose, the gold was more exposed to the elements. Above, the richness is less, but it is easy to access, and the higher the gold, the lesser the quality but easier to extract, because they extract through panning and by what they call here, "guacheo"…Marmato ends up being in the middle of the mountain. This part above is where the gold is less fine, but easier to extract. This is what the state left for us to work.[5]

This distribution of territory and a mineral resource has structured a construction of territoriality that in turn has been decisive in the particular identity construction of Marmato as a mining town that distinguishes it from other communities that structure their social and productive relationships around an enclave economy dependent on a large company. The Marmateños, however, have structured their social and productive

5 Interview with don Eulises Lemus, Marmato mineralogist. March 3, 2012.

relations based on the distribution of territory that has allowed for the co-existence, not without conflict of course, between the small and medium-scale mining, resulting, on one hand, in the maintenance of a permanent and sure source of employment in the medium-scale mining company in the lower zone which approximates the technical and productive criteria that a country dedicated to mining expects. But, on the other hand, this model of territorial distribution has allowed that the inhabitants do not depend on the company, and through the formation of small-scale mining activities in the upper zone, it is possible that there exist "a labor category, free of state controls, that is based on the Marmateña idea that in this mining town the person who does not have money is someone who does not want to work, and that there is no unemployment or hunger" (González Colonia, 2011, cap. II, 19). This vertical territoriality has allowed there to be less asymmetrical social and productive relationships, which in part explains why Marmato, until recently with the advent of the structural inequity produced by the presence of big mining, has not been affected by the internal armed conflict, in contrast to all of the other mining towns and regions in Colombia.

State influences and the incursion of the multinational companies

Since the 19th century there has been a favorable climate for the presence of foreign mining companies in Marmato; in this period, the mines of Marmato were declared to be special reserves of the nation and the exploitation of the mines was handed over primarily to English companies. At the beginning of the 20th century the administration of the mines were taken over by state entities, and with Law 66 of 1946, and the establishment in Marmato of the dual regimen between small and medium-scale mining, there is a legal recognition of what had been an historical co-existence between these two scales of exploitation.

The final two decades of the 20th century and the first decade of the 21st, there is a return of foreign and investment capital. The first enterprise in 1989, that of Dominguez Saieh (which later became Mineros Nacionales), was a medium-scale mine financed with Colombian investors, and followed to certain extent the prevailing territorial scheme of the co-existence of small and medium-scale mining.

With the arrival of foreign capital in the 1990s and the increased presence of multinational companies interested in exploring the gold reserves and the renegotiation of the mining titles that had been acquired, the historical co-existence between small and medium-scale mining begins to unravel; between 1995 and 1997, Gran Colombia Resource Inc. appears on the scene exploring the mineral reserves in the northern part of the municipality, especially in the Echandia part of Marmato, but due to unmet expectations, the company left. In 1996, Conquistador Mines Ltd., a subsidiary of Corona Goldfields, acquired a considerable quantity of the mines in the upper zone of the Cerro el Burro, but later abandoned the municipality such that the mines reverted to their original owners for violation of the purchase contracts. Between 2005 and 2008, Colombia Goldfields Ltd. began to acquire the titles to the mines in the upper zone of Marmato, as well as the processing plants, and negotiated their sale to Medoro Resources, a Canadian company that in addition to purchasing the titles, also continued the exploration of resources, and bought the mine of Mineros Nacionales, thereby taking possession of virtually all of the mining titles in the municipality. The investment made by Medoro has since passed to Gran Colombia Gold Corporation which, in addition to continued exploration, has begun to implement "social responsibility" projects in the municipality which have created very different social dynamics than what has been traditional in Marmato both between the community and the mining companies, but as well, within the community itself.

The latest multinational companies to enter have, until recently,[6] acted in a different manner than the previous companies, not as a speculative investment venture, but rather with a "definitive" proposal to convert Marmato into a large-scale open pit gold mine or a large-scale mine with such socioenvironmental implications as the relocation of Marmato itself (a national historic site), the unemployment of thousands of informal miners (with agricultural projects offered as an economically questionable alternative), and the establishment of a large-scale mining precedent that would impact socially and environmentally other gold-rich areas in the Caldas region although they present these conclusions in a different manner and with different interpretations (which is part of the strategy). This distinct entrance (as of 2006) indicates a different relationship between Marmato, and the state and multinationals than was typical before, particularly because it represents an abrupt rupture with and subsequent fragmentation of a unique Marmateño historical and socioeconomic identity.

The 1990s marks a key moment in the turn of Colombia toward this development model: economic opening, the dismantling of state companies and protectionist measures for local industry, such that the structural adjustments imposed by neoliberalism contain international commitments that reduce the margin of autonomy to resist this model.

[6] Although the situation is not entirely clear or transparent, Gran Colombia Gold, according to some informants, is planning to sell the project and that this is one of the reasons that they have not presented a pre-feasibility study or a proposal to identify the large-scale project.

The transformation of Marmato's "vital space"

The mention of the CC sentence regarding coastal fishermen is not gratuitous in the sense that the "ser marmateño" has more to do "**with the character of their work**" than with other identity criteria insofar as informal mining in Marmato is an historic socioeconomic activity that has brought together diverse communities and identities in the region (indigenous – Embera –, Afro-Colombian, rural agriculturalist, in addition to that of the traditional miner), but where the principal identity – that of miner – is shared by all groups. The speculative exploratory and acquisition activities since 1989, the now omnipresence of the multinational Gran Colombian Gold in the economic, political, and social spheres of Marmato, as well as the uncertain position of the state regarding the future of mining in the area (large-scale/medium scale, open-pit/deep shaft mining), and few opportunities to use the juridical and social mechanisms of participation, has fragmented this intra and inter-community social context and made it difficult for there to be a consolidated social mobilization against the multinational and for their right to access to their sociohistorically consolidated mining activity. At the same time, the sorts of participatory mechanisms that would enable them to demand an audience through Prior Consultation to allow a degree of self-determination in the type of development they want to pursue and a degree of incidence regarding the exploration and exploitation activities of the multinational company, are not available to them because the decision of the of CC is an case-specific interpretation and not the law of the land, even if the evidence points to a shared "vital space".

The potential violence of this fragmentation of their sociocultural and sociohistorical legacies is represented in a number of ways:

1) Resignification of territory and the economic activity associated with it (small-scale vs. medium-scale or large-scale mining; informal vs. formal; subsistence and/or sustainable development vs. a developmental paradigm)

Law 66 of 1946, created the legal possibility for the co-existence of two different scales of mining, legitimating what had been an historical practice since the 19th century.

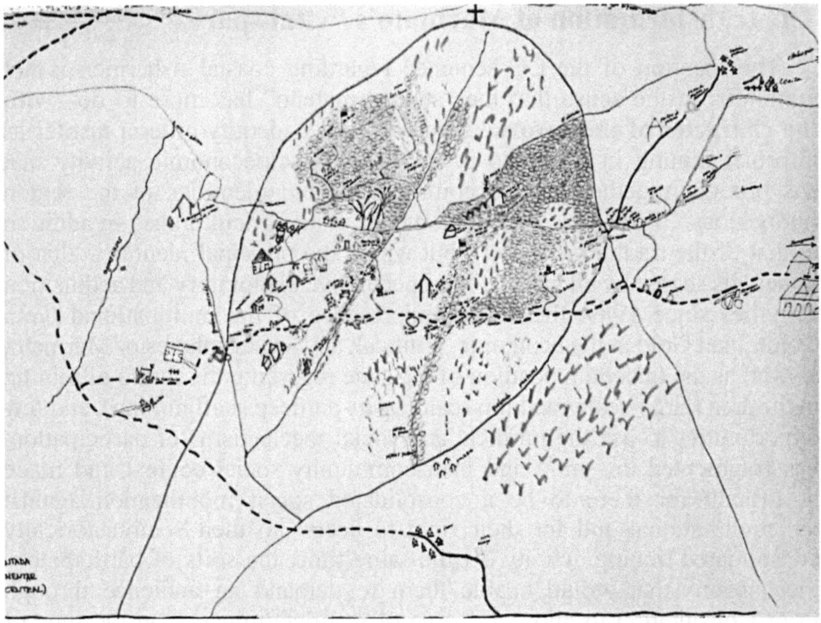

Social cartography of Marmato. Mauricio Castañeda, traditional miner. Marmato. 2012

More than a simple diversification and expansion of economic development models in the region, the law represents a socioeconomic division and epistemological territorialization of what is socially appropriate economic activity with all of the social practice and social relations, etc., that accompanies these activities. Even more so, there is an implicit designation of traditional mining as something outside of the legal parameters of "modern" mining. The informality of labor relations and contractual obligations, a different work ethic and *raison d'être* for subsistence mining, the lack of health and social protection, the lack of technification, the precarious relations with the state, are indicators used by the state, the multinational companies, and the NGOs that accompany the social work of the mining companies, to qualify the informal mining as outside of the operative development model.

Informal mining is a socioeconomic tradition that, rather than representing a legitimate alternative to the predominant development model, is thought to act as an obstacle presented by people not recognized as distinct to the mainstream population or to the national identity (with the exception of organized indigenous and Afro-Colombian communities). What is more, there is an incomprehension on the part of the hegemonic actors – state, multinational – as to why the Marmateños would be against the sort of development model they are proposing; to a large extent, the opposition that the miners are presenting – they are not against mining per se, just the type of mine that is being proposed – represents a perspective the multinational is not necessarily prepared to understand or address. If the issues were a question of the environmental impact, or of protected identities such as ethnic ones, the responses and policies are fairly standard. But the question represented by the Marmato case of how to mine and not whether to mine, and the reasons behind a type of mining that is not based on the accumulation of capital, are questions to which the multinationals cannot respond.

And what is seen as the socioeconomic improvement of a specific population through the development initiative offered by the proposed large-scale (open-pit) mine threatens this socially significant vital space.

2) A consolidated historical and socioeconomic identity transforming into a politicized ethnic one, and the invisibilization of the traditional miner

There is a consensus among Marmateños that although there was and is recognition of the different ethnic groups – indigenous Embera and Afro-Colombian – in Marmato, the differences these identities might represent did not have much impact on the social relations and social identities that were primarily based on the artisanal mining tradition. To a certain extent, these identities are territorially based; the Embera are mostly in the San Juan sector of Marmato and the Afro-Colombians are largely represented in El Llano. But these represent historical residential tendencies instead of clearly defined ethnic spaces; the mines they work are not so spatially defined in terms of ethnicity.

Regarding the emergence of ethnicity as a factor to consider in Marmato, there are several issues involved: first, the significance of ethnicity as a primary identity is largely the result of external factors and outside agency than it is to some internal need to foreground ethnicity; second, the issue of Prior Consultation and the limitations this represents (until the recent CC decision) to non-ethnic populations, and, third, the foregrounding of ethnicity in the Political Constitution of 1991, and the subsequent political privileging of ethnic identity, suggest that in circumstances that threaten the local populations in Marmato, there is a political expedience to emphasize ethnicity over other local, historical constructions of identity. In this regard, the ethnic communities of Marmato, especially the indigenous Embera, aided by CRIDEC[7] and other indigenous organizations, began to organize

[7] The Consejo Regional Indígena de Caldas (CRIDEC) is a second-level organization, founded in 1985, that groups together all of the indigenous community councils ("cabildos") in the Department of Caldas. It has played an active role in the processes of the reconstruction of indigenous identity (or reindigenization) that are taking place in the region from the 1990s to the present.

themselves on the basis of different social and identity criteria than had been traditional in Marmato, with the potential to begin to articulate their recognition as ethnic communities by the state and the multinational company, and the issue of the impact of the large-scale mining project.

The centrality of the mining tradition among the different populations of Marmato as a source of identity is still evident, but the emergence of other mitigating identities such as ethnicity are beginning to take hold especially due to the differentiated rights accorded the different populations and how these rights turn into differentiated strategies to protect traditional mining and the social context of this economic activity. It is an ambiguous situation for a number of reasons. First, established social relations are threatened, and the sorts of social and solidarity mechanisms that have defined Marmato as a unique context are changing in ways that are not coherent with Marmato. Second, what is operative here is a legal and social presumption of legitimate and natural collectivities represented in ethnic groups that is not granted to other forms of collective identities, even if there is a demonstrably distinct socioeconomic epistemology. And third, even within recognized ethnic groups, the criteria to represent those groups are no longer consensual because of third party influences. One of the unfortunate consequences was the gradual adherence to the external social and state criteria of *otherness* which began to change local identity narratives both within the different ethnic groups and between them and the non-ethnic population reifying the near and distant *other* distinction against the traditional identity marker.

In Marmato, there were significant differences between previous indigenous community leaders and the current one that were taken advantage of by the multinational. The differences were based on what being indigenous means, and how do you guarantee well-being for the community. A former leader of the community who first organized the indigenous council that governs the community understands the vindication of indigenous identity as a process of becoming part of modernity, of being able to access basic services and goods such as healthcare, education, and family subsidies, that have been out of reach to the community, and understands that the defense of the community in the context of a large-scale mining project is to negotiate the conditions of the community's inclusion in the project: employment, works that benefit the community, etc. The current leader (2011-2013) understands indigenous identity as the defense of territory, which in this context implies that any mining project that threatens small-scale traditional mining also threatens indigenous identity.

In terms of socializing the proposed mining project, negotiating their social responsibility, and social investment, Gran Colombia Gold made contact with previous leaders not recognizing the legitimacy of the current leader and her vocal opposition to the mining project, thereby undermining her legitimacy, if not her authority, in the community and among other actors in Marmato, and creating other fragmentations in recently consolidated identities.

3) Circumstantial (closing the mines) reclassification of traditional economic activities as illegal (guacheros), and the imposition of ideas of development

Traditionally in Marmato, the right to exploit a small-scale mine has not been tied to a legal title or license to mine as is the case in other parts of Colombia.[8] The association of one or more persons to "work" in closed or abandoned mines is recognized as one of the strategies that has permitted the conformation of small and medium-scale mining companies in which a considerable majority of the local and migrant population is employed; mines that, in some cases, have managed to legalize their titles according to Colombian law or that are in a decade-long process of legalization.

The arrival of Colombia Goldfields in 2005 to Marmato introduced a major change in the exploitation dynamics. Initially the company negotiated the cession of the titles that were in the hands of the local miners. Subsequently, it bought the "molinos" used to process the material. The first action implied the closing of mines and the firing of the local miners, who either remained unemployed or migrated to other regions of the country. The purchase of the "molinos" was followed by their destruction on the part of the mining company which the president of Colombia Goldfields justified suggesting that is was a necessary measure to improve the quality of the streams and water affected by the processing of the extracted material.

---

8 For many years (1940-1980) the mines were under the direct administration of the state (through the Direcciòn de Minas de Marmato, that operated directly in the municipality), which explains why the relation between the miners of the municipality and the mining authority was not mediated in many cases by written titles rather by verbal agreements. From the 1980s, the area of Marmato was constituted as a "support" ("aporte") (a figure that handed over to a state business the right to explore an area or to contract that exploration to third parties) that was administered by a succession of state businesses (Ecominas, Mineralco, and Minercol) that operated outside of the municipality and periodically sent officials to supervise the mining area. During this period, the informal aspects of mining were maintained, not so much because of the distance of the mining authority, rather for the lack of control on the part of the state enterprises charged with the exploitation.

"Molino" with a "mesa alemán" in Echandia (Photo courtesy of Paula Hinestroza, 2012).

The actions on the part of Colombia Goldfields affected in a direct manner the local social and economic customs and augured the arrival of the recharacterization of a local mining figure, the "guachero". The "guacha", the local name for this type of activity, has been an historically viable economic alternative when independent miners reopened abandoned mines to practice a subsistence form of independent mining exploitation. But in the present context, due to the closing of the mines by the company, unemployed miners began to enter and exploit the closed mines at night as a way of responding to the employment crisis and the lack of a response on the part of the state to the activities of Colombia Goldfields.

Nevertheless, with the recent intensification of the state discourse regarding illegal mining activities, in conjunction with the prioritization of large-scale mining as one of the principle development initiatives, and the imposition of administrative protections ("amparos administrativos")[9] on the part of the multinational companies that have acquired titles to the mines in Marmato with the objective of removing the *guachero* miners in the recently acquired titled areas. A traditional and subsistence practice is converted into an illegal one subject to sanctions and penalization according

9 Legal institution that allows a mining title-holder to obtain legal protection from those who exploit the mine without a title.

to the juridical dispositions and the regulation of mining activities. One important aspect is the relationship that has been established between illegal mining and the criminal mining – this last associated with illegal armed groups – with the indiscriminate use of these terms on the part of the state's representatives and the mass media. This fact results in the erasing of the characteristics and of the socioeconomic implications of each one of these practices, and to a certain extent affects the visibility of the question of tradition as the basis of the configuration of "vital spaces" in places such as Marmato (See Giraldo y Muñoz Mora, 2012).

Currently, the state is promoting the formalization of small-scale mining as a form of control over an economic activity that has not generated the sort of economic and social development that the state seeks to develop. It concerns a policy supported by the multinational companies and attempts to locate traditional mining within the logic of the capitalism that motivates large-scale companies, using such terminology as "more efficient", "greater investment", "technical and business formalization", "adequate technical standards", among others. In this manner, small-scale mining represents the first step in the development of a mining industry, which must be regulated to achieve its maximum potential, at the same time that this development ignores or discredits the socioeconomic logic that motivates small-scale mining, as in the case of the guacheros of Marmato: a principally subsistence activity, autonomous, where the measure of success is its permanence in time – the possibility of always having access to a resource in the Cerro el Burro – and not in terms of greater productivity to maximize immediate profits.

4) The threat of relocation, and the actual process of relocating community institutions (hospital, school)

Relocation implies the sacrifice not only of properties but also of valued social spaces which means that affected populations must begin a process of the "reconstruction of their cultural and productive environment at the same that they must maintain their modes of production" (Robinson, 1998, p. 248). This concerns affectations that are continually disregarded in development contexts with the justification that this represents "inevitable social costs".

Actions are being taken to make the upper part of Marmato untenable by reducing opportunities for traditional economic activity, and transferring basic services to El Llano or the lower part of the municipality, and the current relocation destination in the case that the option for an open-pit mine is acted on.

The relocation of the municipal center of Marmato to the El Llano section has been a constant rumor[10] in the recent history, and justified by reference to the geological risks that have resulted from the small-scale mining activities in the municipal center and the Cerro el Burro that rises behind it; in 2006, a flow of mud from above flowed into what was the central plaza destroying houses and business, and forcing the subsequent relocation of the Banco Agrario and Municipal police station to El Llano, the hospital to a series of houses (a temporary solution while they built a new hospital en El Llano), and the abandoning of the central plaza, and the surrounding houses and businesses.

Nevertheless, the arrival of Colombia Goldfields in 2005 and its marked interest in an open-pit mine, increased the rumors and the perception that the relocation option was the Colombian state's direct support of the multinational's interests and an imminent threat to the social and economic forms of the local populations, a feeling that was exacerbated with the arrival of Gran Colombia Gold in 2011, and its direct financial support toward the construction of the new hospital and high school in El Llano.

In this context, opposition to the relocation of the municipal center is unanimous and constructed around the idea of Marmato as a place[11] configured, on one hand, around a particular economic activity that has structured socioeconomic relations, cultural practices, a sense of territoriality and landscape, and without which it would be impossible to reconstruct their lifeways, but fragmented, on the other hand, into different interests, leaderships, identities, and positions regarding the multinational.

For this reason, the different public infrastructure projects that have been started in El Llano to date are looked upon with a certain trepidation by the local population in the upper zone and municipal center, first because they consider that the decentralization of public services creates barriers to access the same for the population that is located in the sector referred to as "El norte", which is the sector most distant from the municipal center and significantly more so from El Llano, which means more time and greater cost to access the services.

---

10 This rumor has been accompanied by concrete actions: the almost total relocation of the principal state institutions from the municipal center to El Llano, and choosing El Llano as the area of municipal expansion in the Scheme for Territorial Reorganization (EOT), which is to be approved by the municipal council. Alleging geological risks, the EOT contemplates the slow but complete relocation of the municipal center.

11 Here we assume the idea of Escobar to approach the place in "terms of the individual entities and constituted historical processes that operating at different levels, interact in complex forms to conform a totality" (Escobar, 2011, p. 78); in this way, and following the idea of the same author, the place refers to "a particular location with a certain measure of rooting (even if unstable), some limits (that are permeable), and some connections to quotidian life, even when its identity is constructed and never fixed" (p. 47).

As well, they consider that the imminent transfer of the high school, hospital and other administrative services to El Llano will result in the decline of the municipal center as the locus of commercial and administrative services; a situation that favors the interests of the multinational companies that want to extract gold in large scale, reducing even more the possibility of inhabiting this part of the municipality. This concerns a fear felt not only by the inhabitants of the municipal center, but by all of the sectors of the municipality, including El Llano, because of its long-term impact:

> Investigator: And you, as inhabitants of El Llano, aren't you worried that people will start to arrive, people to live here?
>
> FV: Do you mean the people from above?
>
> Investigator: Yes.
>
> FV: But what could be worrisome? Leave that space up there alone, because if it is left uninhabited, then there is nothing, nothing. But there are a lot of people who are not aware that if it begins up there, after that it will continue down here [in El Llano].[12]

5) Social projects and investments by the multinational companies that reinvent leaderships, privilege conformity to the multinational project, serves individual interests over collective ones

One of the aspects that was hardest to demonstrate was the disinterested manner in which the multinational operated in spite of their manifestations of largesse toward the community of Marmato. The Director of Sustainability of Gran Colombia Gold, the person in charge of community relations and corporate social responsibility, offered some suggestive ideas as to how the company perceived the community, and, more importantly, how it considered its representations to be more legitimate than local community ones.

> With the indigenous community from the first moment it was impossible to approach the leader (gobernadora)... because she presented a total opposition [to the project]. Dialogue was impossible. I was forced to invite a former traditional leader ... who introduced me to the other ex-traditional leaders. We wrote a letter of intent that at this moment is being reviewed by CRIDEC.
>
> One of the criteria was to identify the *true* leaders because one of the errors that many people commit is to not talk with the community but rather with the opposition. There are leaders who are part of Pro Defensa,[13] but I don't see the ones who truly want the [multinational mining] company to arrive being interviewed.

---

12 Testimony of a leader of the Asociación de Barequeros de Marmato. March 2012.

13 Comité Civico Pro-Defensa de Marmato is an organization that formed in 2005, which groups different social organizations and persons in the municipality who are opposed to the open-pit mining project.

> If we could only be an example of how a multinational could bring development to a community.
>
> There are some persons with an institutional affiliation, but who do not represent the people. Pro Defensa de Marmato doesn't represent, not 10%, not 15%, of the community. But I have an open forum in "pro defense" of Marmato.
>
> These people meet to protest against the company, but I am constructing their community center for them, because we have to support not only those who are in favor of the company, but as well those who are against it.
>
> For this reason, one of the fundamental tools is to strengthen the leadership of the community, so that the community has leaders.[14]

The strategy of the company is to resignify materially and discursively what is Marmato's "vital space", replace and replenish through community-based programs a compliant leadership, and empower sectors that have played a subordinate role in the community even as they contribute to the historical identity constructions: traditional Marmato is in need of the company's vision of its future, its leadership insofar as it manifests dissidence to the prevailing development scheme is not a leadership and does not represent the community, it is within the role of the company to identify what is the true leadership in the community and to only work with them in detriment of community-recognized leaders, and, finally, the company can buy allegiances through community actions and the empowerment of counter-perspectives to the prevailing opposition to the large scale mining project.

The only recourse available to the people of Marmato and only to those who possessed actual title to their mines was to refuse to cede their mining rights to the multinational companies. But this recourse was not used due in part to the fact that the majority saw it as an economic issue that privileged individual interests over the future consequences of the development project.

Once the multinational companies acquired the mining titles, there remained few recourses, legal or otherwise, available to them except civil disobedience ("vías de hecho"), in part because only certain members of Marmato can appeal to the special rights accorded to ethnic groups, and any decision would only apply to their ethnically identifiable territorial space, and not to the socioeconomically-based vital space that encompasses all of Marmato and the different groups and families that constitute its social fabric, and in part due to the increasing incidence of external actors – state, multinationals, NGOs, armed groups, etc. – that do not recognize or

---

14 Interview with Luca Altamura, Director of Sustainability of Gran Colombia Gold. April 17, 2012.

understand how Marmateños interpret their territoriality, their vital social space.

No one pretends to suggest that there has not been internal conflict in Marmato, or that the community divides on any number of issues, this is part of being a community and particularly one that has constructed a unique socioeconomic profile. Like the coastal fishermen in the CC decision, the Marmateños want to participate, but on their own local terms, in recognition of their uniqueness, and protecting not only their livelihood, but as well their historically constructed identity. One of the mine owners who has played an important role in the development of small-scale mining suggests:

> Here there is no opposition to the arrival of the multinational. Here the opposition is to the open-pit mine. If a multinational arrives here and proposes a deep-shaft mine, the people would be willing to stay. **The tranquility here is based on conserving this.** We know that if they start an open-pit mine, in 30 years there won't be anything left.[15]

Some people suggest that while most of the Marmateños are concerned with the possible open-pit mining project and have armed their opposition specifically against the idea, the true mining project is different, and in fact has already begun:

> One doesn't see an open-pit culture, rather that of deep-shaft mining beginning with Mineros Nacionales and consuming the mine upwards. We are all concentrating in what is happening in the upper zone, but the true exploitation is below. But it is impossible to see what is happening. It is like discussing with someone about the characteristics of a banana. While one person discusses it, the other eats the banana. In Marmato, there is a serious enterprise called Mineros Nacionales, and not just for a day, they use a lot of explosives. I think what is happening is the same as in the movie "*La estrategia del caracol*", such that while they are arming the legal battle [to conserve the upper zone], they are taking away the house from below. With the current price of gold, subterranean exploitation is very profitable.[16]

By way of conclusion

The recent state characterization of informal mining as illegal has placed the burden of legitimacy on the traditional miners of Marmato and other sites of small-scale mining, and this in a context where the very same state has taken away the legal resources available to them that would validate their centuries' old construction of a viable mining community and of a

15 Interview with Alberto Valencia. March 22, 2012. Our emphasis.

16 Interview with Alexander Restrepo, expert lawyer in Mining Law. April 9, 2012.

mining identity that has until recently transcended other identity markers such as ethnicity.

The case of Marmato shows the manner in which the development discourse becomes part of the local conversation with the arrival of a large-scale mining project in a region where small and medium scale mining has traditionally co-existed. This intrusion implies a series of transformation in local identities – the reification of the near and distant *other* – the construction of territory and the deterritorialization of its inhabitants, and the actions of the local population oriented towards either adaptation to, or, to the contrary, resistance to – through legal means or civil disobedience (legislated and non-legislated mechanisms of participation) – the new actors, summarized in the term "multinational", that contest local control of territory and its resources.

One of the last episodes of this case that occurred after the fieldwork took place in August 2013, during the national strike conducted by small and medium-scale miners regarding their displacement from their sites of traditional mining, the criminalization of traditional and informal small-scale mining, and the arrival of big mining to Colombia. In stark contrast to a daily life that was characterized by the lack of armed violence, during the strike the people of Marmato lived through one of their more difficult moments in the confrontations between armed public law enforcement that repressed the striking miners who defended their way of life with rocks and shovels. This episode of the *illegalization of traditional mining* shows that the epistemic violence that defines the state's and industry's mis-characterization of traditional mining transcends the symbolic level to resort, finally, to physical violence.

Bibliography

Colombia, Asamblea Nacional Constituyente. (1991). Constitución Política de Colombia.

—, Congreso de la República. (1946). Ley 66 de 1946, por la cual se dictan disposiciones sobre las minas de Marmato.

—, Congreso de la República. (1991). Ley 21, por medio de la cual se aprueba el Convenio número 169 sobre pueblos indígenas y tribales en países independientes, adoptado por la 76a. reunión de la Conferencia General de la O.I.T., Ginebra 1989.

—, Corte Constitucional. (2009). Sentencia T – 769 de 2009, M.P.: Nilson Pinilla Pinilla, Bogotá.

— (2012). Sentencia T-348/12 del 2012, M.P.: Jorge Ignacio Pretelt Chaljub, Bogotá.

—, Presidencia de la República. (1954). Decreto 2223 de 1954, por el cual se dictan algunas normas para la contratación de las minas nacionales de Marmato.

Dover, R., *et al.* (mss.). *Inclusiones incompletas: desarrollo, participación ciudadana y Consulta Previa.* Manuscript sent for consideration to publish to the Instituto Colombiano de Antropología e Historia.

Escobar, A. (2010). Antropología y desarrollo. En Guillermo Julián Hernández, (ed.), *Antropología y desarrollo. Encuentros y desencuentros.* La Habana: Centro Nacional de Superación para la Cultura, Colección Punto de Partida. pp. 29-58.

—, (2011). *Territorios de diferencia; lugar. Movimientos, vida, redes.* Popayán, Colombia: Envión Editores.

Giraldo Ramirez, J. y Muñoz Mora, J.J. (2012). *Informalidad e ilegalidad en la explotación del ORO y la MADERA en Antioquia.* Medellín, Colombia: EAFIT y Proantioquia.

González Colonia, C.J. (2012). *Brujería, minería tradicional y capitalismo trasnacional en los Andes colombianos. El caso del pueblo minero de Marmato, Colombia.* (Tesis de maestría inédita). Universidad Nacional de San Martín, Buenos Aires.

Gupta, A., and Ferguson, J. (1992). Beyond "Culture": Space, Identity, and the Politics of Difference, *Cultural Anthropology*, 7(1): 6-23.

Hale, C. R. (2005). "Neoliberal Multiculturalism: The Remaking of Cultural Rights and Racial Dominance in Central America", *Political and Legal AnthropologyReview (PoLAR)*, 28 (1), pp. 10-28.

Lopera, G. y Dover, R. (2013). "Consulta Previa, ciudadanías diferenciadas y conflicto socioambiental", *Boletín de Antropología* 28(45): 76-103.

Lasso Lozano, M. (2012). *Campo de la minería de oro y habitus productivo en Marmato, Caldas: estrategias de conservación y transformación en tiempos de globalización. Marmato 2001 – 2011.* (Tesis inédita de pregrado), Universidad de Caldas, Manizales.

Ramírez, O. (2010). "El espejo invertido de la realidad: del discurso del desarrollo a la apología de la gestión ambiental". In Guillermo Julián Hernández (ed.), *Antropología y desarrollo. Encuentros y desencuentros.* La Habana: Centro Nacional de Superación para la Cultura, Colección Punto de Partida. pp. 79-90.

Robinson, S.S. (1998). "Desarrollo, relocalizaciones involuntarias y redes de informática rural: múltiples escenarios, guiones y el perfil de un contexto emergente". In S. Zendejas y P. de Vries (eds.), *Las disputas por el México rural* (pp. 243-254). Michoacán, México: El Colegio de Michoacán.

Scott, J. C. (1998). *Seeing like a state: how certain schemes to improve the human condition have failed.* New Haven: Yale University Press.

Analyse comparée du phénomène de la violence contre les migrants au Maroc et au Mexique, pris dans le contexte des processus d'intégration régionale

Émeline Nanga

Chercheure en formation doctorale, Université du Luxembourg, Unité de Recherche Identités. Politiques, Sociétés, Espaces (IPSE)

I. Introduction

Cet article s'interroge sur les processus ainsi que sur la dynamique des formes de violences et d'exploitation auxquels sont confrontés les migrants irréguliers au cours de leur trajet migratoire vers et à travers le Maroc et le Mexique dans l'objectif final de se rendre respectivement aux États-Unis ou dans l'UE, spécifiquement au regard des processus d'intégration en cours dans ces espaces. L'objectif pour nous étant de saisir la nature exacte du phénomène de la violence dont sont victimes les migrants dans ces espaces, ainsi que la manière dont il sied d'évaluer les agents ainsi que les facteurs responsables de celle-ci. Tout en essayant de mettre en relief la nature des liens qui existent entre les processus d'intégration économique régionale, les stratégies de développement, les régimes de sécurité régionale et les contrôles migratoires, ainsi que leurs incidences sur les droits des migrants vivant ou transitant dans les espaces concernés.

Nous prendrons spécifiquement en compte la dynamique des violences qui affectent les migrants irréguliers originaires d'Afrique subsaharienne et d'Amérique latine (Amérique centrale/Sud) dans le cadre de leur périple migratoire vers et à travers le Maroc et le Mexique, respectivement en direction de l'Europe et des États-Unis. À cet effet, nous aurons recours aux concepts de Galtung *et al.*, développés sur la violence directe, structurelle et culturelle, que nous jugeons assez pertinents pour comprendre et contextualiser la nature des violences qui affectent les migrants lors du franchissement irrégulier des frontières dans les espaces étudiés ; et par ailleurs, mieux adaptés pour appréhender et illustrer tant bien leur caractère récurent, voir systématique, ainsi que l'impunité qui les caractérise. La problématique ainsi dégagée étant celle de savoir quels sont les types de violence qui ont lieu dans ces espaces ? Comment cette violence se reproduit-elle ? Et de quelle manière elle affecte les migrants ?

Notre hypothèse repose sur l'idée selon laquelle l'antagonisme entre les stratégies de libre-échange, auxquels participent les gouvernements indexés et les politiques migratoires restrictives appliquées aux frontières des pays concernés, crée des agendas politiques contradictoires au niveau de la prise des décisions ; avec d'une part des actions qui soutiennent les migrations économiques par la création des marchés régionalisés, et d'autres qui les restreignent à travers le renforcement des mesures contrôles aux frontières nationales. Cette incohérence politique favorise la violation des droits des migrants en ceci qu'elle participe à l'accentuation du caractère « informel » de la migration et à l'aggravation de la vulnérabilité des migrants face à la violence et aux abus divers, y compris le trafic et la traite des êtres humains.

Cette communication s'articule autour de cinq sections principales, inclue la présente introduction. La deuxième section portera sur la conceptualisation et la transposition de la violence dans le champ des études migratoires. La troisième et la quatrième section seront respectivement consacrées à la violence contre les migrants au Mexique et au Maroc. Tandis que la cinquième section qui fait office de conclusion, visera à souligner la nature des liens entre la violence contre les migrants et les régimes de sécurité régionale. Fondée sur une démarche qualitative, notre argumentation s'appuiera principalement sur les sources primaires et secondaires recueillies dans la littérature générale sur la migration irrégulière au Maroc et au Mexique, les textes formels des gouvernements concernés se rapportant spécifiquement à cette question, et les rapports spéciaux établis par la société civile œuvrant dans le champ migratoire.

II. Conceptualisation de la violence et sa transposition dans le champs des études migratoires

Depuis toujours, la mobilité humaine a été liée aux processus de violence, à l'image de la traite négrière transatlantique ou des demandeurs d'asile fuyant les génocides et la guerre. Les effets du contrôle, de la militarisation et du racisme étant particulièrement saillants dans les espaces migratoires situés dans les régions frontalières, les zones industrielles clés et les enclaves. Cependant, peu d'études se rapportent à l'espace physique dans lequel se déploie le parcours des migrants, et la manière dont celui-ci s'est transformé en un champ de violence. Ceci étant particulièrement vrai pour les sans-papiers tournés vers des filières clandestines, et qui sont nombreux à effectuer des voyages dangereux et coûteux autant en termes de vies, que de temps et d'argent. Le taux inquiétant de tragédies et de pertes humaines recensé le long des principales voies empruntées par les migrants, à l'instar des migrants subsahariens au Maroc et ceux originaires d'Amérique centrale et du Sud au Mexique, illustre parfaitement cette réalité brutale.

Afin de mieux saisir et contextualiser la réalité des violences vécues par les migrants clandestins qui vivent ou transitent vers et à travers le Mexique et le Maroc dans l'objectif final de se rendre respectivement aux États-Unis ou dans l'UE, nous suggérons d'avoir recours aux concepts de violence structurelle, directe et culturelle élaborés par Johan Galtung *et al.*, dont les champs de définition couvrent aussi bien les actes d'agression physiques que symboliques, ainsi que ceux susceptibles de priver les individus des éléments vitaux essentiels à leur bien-être, leur santé ou à leur développement, indépendamment des agents ou des facteurs structurels jugés responsables de cette situation.

Conceptualisation de la violence selon Galtung et al.

Johan Galtung, plus connu pour ses recherches sur la paix, est l'auteur d'une des théories les plus complexes sur la violence. Selon Galtung (1969 : 168), « *Violence is present when human beings are being influenced so that their actual somatic and mental realizations are below their potential realizations* » [« la violence est présente lorsque les êtres humains sont influencés de telle sorte que leurs relations effectives, somatiques et mentales, se situent en dessous de leurs réalisations potentielles »]. Afin de déterminer les responsabilités, Galtung (1969 : 170-171) va segmenter ce portrait général de la violence en trois concepts distincts représentés sous la forme d'un triangle dont : (1) *la violence personnelle ou directe*, (2) *la violence structurelle ou indirecte* et, (3) *la violence culturelle.*

Figure 1 : Typologie des violences selon Galtung

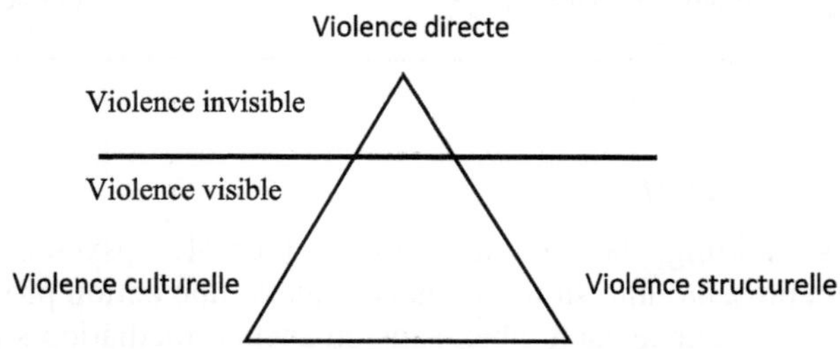

Source : J. Galtung (1998). *Tras la violencia 3R : reconstrucción, reconciliación, resolución. Afrontando los efectos visibles e invisibles de la guerra y la violencia.* Bilbao : bakeaz, gernika gogoratuz. p. 15.

Dans son triangle, Glatung démontre qu'il existe une relation temporelle entre ces trois concepts de violence. La violence directe étant considérée comme un événement ; la violence structurelle comme un processus en dents de scie, et la violence culturelle comme une constante fondamentale qui reste inchangée sur de longues années en raison des transformations lentes de la culture de base (Galtung, 1999 : 9 ; Galtung, 2003 : 11). Il opère un lien direct entre la violence culturelle et la violence structurelle. Les trois types de violence étant liés en ce sens que la violence directe peut être un indicateur du degré de violence structurelle et culturelle dans la société (Galtung, 2003 : 13 ; Penalva et La Parra, 2008). La violence directe étant considérée dans ce sens comme la pointe émergée d'un iceberg et la violence culturelle, comme le pilier qui génère et consolide la violence structurelle et dont les effets s'étendent à long terme, comparativement aux autres formes de violence.

Figure 2 : Différenciations de la violence directe et indirecte, selon Galtung

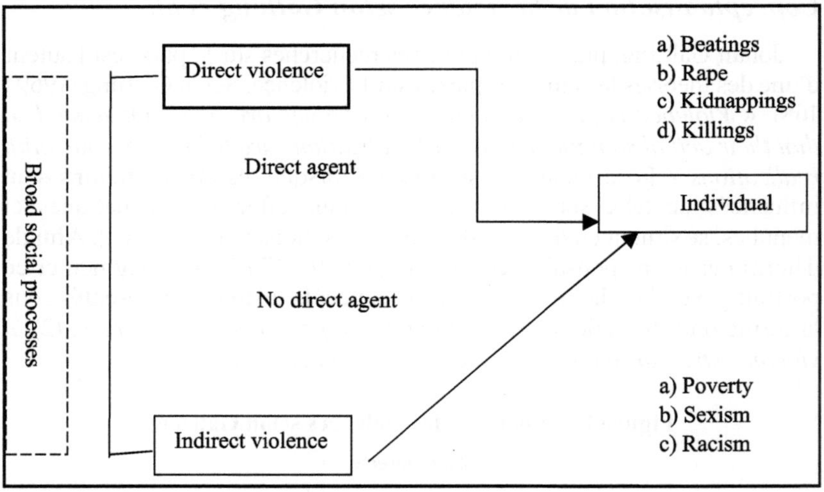

Source : J. Galtung (1998). *Ibid.*, p. 15.

La violence directe

D'après Galtung, la violence directe (verbale, psychologique et physique) représente une situation dans laquelle une action provoque un dommage direct sur le sujet cible, sans qu'aucune médiation s'interpose entre le début et la fin de celle-ci. Il s'agit d'une relation violente entre les entités humaines (personnes, groupes, ethnies, les institutions, les États, les coalitions), suivant le processus sujet-action-objet, sans que rien ne s'oppose quasiment à sa mise en œuvre. Par définition, la violence directe est visible, destructrice et perpétrée dans l'intention manifeste de nuire, et constitue de ce fait la forme de violence la plus redoutée.

Ses effets étant souvent irréversibles, ou ayant parfois des répercussions psychiques majeures. Le terme de violence directe est généralement utilisé pour la distinguer d'autres formes de violence (structurelle, culturelle et/ou symbolique), dans laquelle la conséquence des actions privent les individus des éléments essentiels à leur bien être par le biais d'un processus continu dans le cadre duquel les décisions sont prises avec un certain recul, marquées par les acteurs sociaux qui les exécutent, au cours de la période dans laquelle l'action s'écoule et dans l'espace ou celle-ci survient. Ainsi, la violence directe constitue-t-elle une analyse des privations des besoins fondamentaux nécessaires à la survie d'un individu, à la garantie de sa sécurité, de son bien-être, son identité ou de sa liberté. Ses conséquences entrainent généralement la mort (ex : homicides, les génocides ou les assassinats), la misère, le chaos, la relégation au rang de citoyen de seconde classe, la resocialisation ou la désocialisation à sa propre culture, la répression, les arrestations, les expulsions etc.). Autant de faits qui permettent de saisir à quel point les besoins fondamentaux des individus sont perceptibles dans le champ de la violence.

La violence structurelle

Galtung considère la violence structurelle (ou indirecte) comme un concept plus large dont les manifestations s'étendent au-delà de la simple violence physique. Le concept de la violence structurelle se concentre explicitement sur l'analyse des inégalités et des mécanismes sociaux, politiques et économiques employés pour la créer ou la renforcer, et maintenir ainsi la marginalisation continue d'un groupe de personnes (Galtung, 1969 ; Nevins, 2005 ; Farmer 2004, Jiménez et Muñoz, 2004, Kleinman 2000). Galtung estime que la violence structurelle fait partie de la structure et que sa caractéristique première est l'inégalité. Il s'agit d'une violence institutionnalisée et validée par les structures sociales et de pouvoir, et en rapport avec des faits tels que l'injustice structurelle sociale, économique et éducative, l'inégalité des chances dans le développement humain, la pauvreté et l'aliénation, la marginalisation et la violence politique etc. qui, associée à la violence directe, affectent les individus dans leur plus grand nombre et accentuent leurs besoins fondamentaux, tout en générant des nouveaux cycles de violences (Galtung, 1995, Galtung, 1985 : 31-96). Cette conceptualisation large de la violence s'avère intéressante en ceci qu'elle nous éclaire sur le caractère insidieux de la violence ainsi que sur ses modalités qui nécessairement ne sont pas visibles à l'œil nu.

La violence culturelle

La violence culturelle est perçue par Galtung (1990 : 291) relevant de certains « aspects de la culture » (et appartenant à la sphère symbolique de notre existence telle que la religion, l'idéologie, la langue et l'art, la science empirique et formelle), qui peuvent être utilisés pour justifier ou légitimer la violence directe ou structurelle contre un individu ou groupe d'individus (ici, les migrants). La définition de la violence culturelle telle que formulée par Galtung, s'accorde également avec le concept de Bourdieu (1977 : 191) sur la violence symbolique, en référence à la stigmatisation des certaines pratiques. On peut faire référence dans ce sens à la stigmatisation des pratiques migratoires, si l'on considère que la façon dont un individu se perçoit peut être affectée par la discrimination et le dénigrement. Autrement dit, la violence symbolique ne sert pas seulement à légitimer, mais parfois à masquer d'autres formes de violences, en attribuant la responsabilité de ces faits à des sources de vérité distinctes (voir aussi Imbusch, 2003 ; E. Terray, 1996). Ainsi la violence symbolique peut-elle dans certains cas, constituer la face cachée de la violence structurelle.

Transposition du concept de la violence dans champ migratoire

En tant que processus social, la migration s'imprègne dans la violence et l'insécurité. Partant des facteurs structurels à l'origine du phénomène migratoire dans les pays d'origine des migrants, à la vulnérabilité constante à laquelle ils sont exposés durant l'ensemble du processus migratoire, principalement dans les pays d'accueil et de transit. Ainsi peut-on voir que, la typologie des violences établie par Galtung *et al.*, s'applique parfaitement dans le cadre du franchissement des frontières, ainsi que dans les différents contextes liés à celui-ci. Cette conceptualisation de la violence s'avère notamment essentielle dans la saisie de la complexité qui caractérise les violences subies par les migrants durant l'ensemble de leur périple migratoire, de leur pays d'origine aux pays de transit et d'accueil. Elle permet également de saisir le caractère prévalent de cette violence, ainsi que la systématisation de celle-ci. Sans oublier de souligner le climat d'impunité qui entoure ces faits, ainsi que les forces sociales, politiques et institutionnelles qui la perpétuent. Les abus ainsi que les souffrances multiples dont sont victimes les migrants (irréguliers) au cours de leur parcours migratoire étant marqués à différents niveaux, par des faits de violence majeure aussi bien d'ordre direct que structurel et culturel (cf. figures 2 et 3) ; alimentées par le climat d'impunité général qui prévaut dans ces contextes (cf. figure 4).

Figure 3 : Violence directe et indirecte durant le transit : Violences directes et indirectes durant le transit

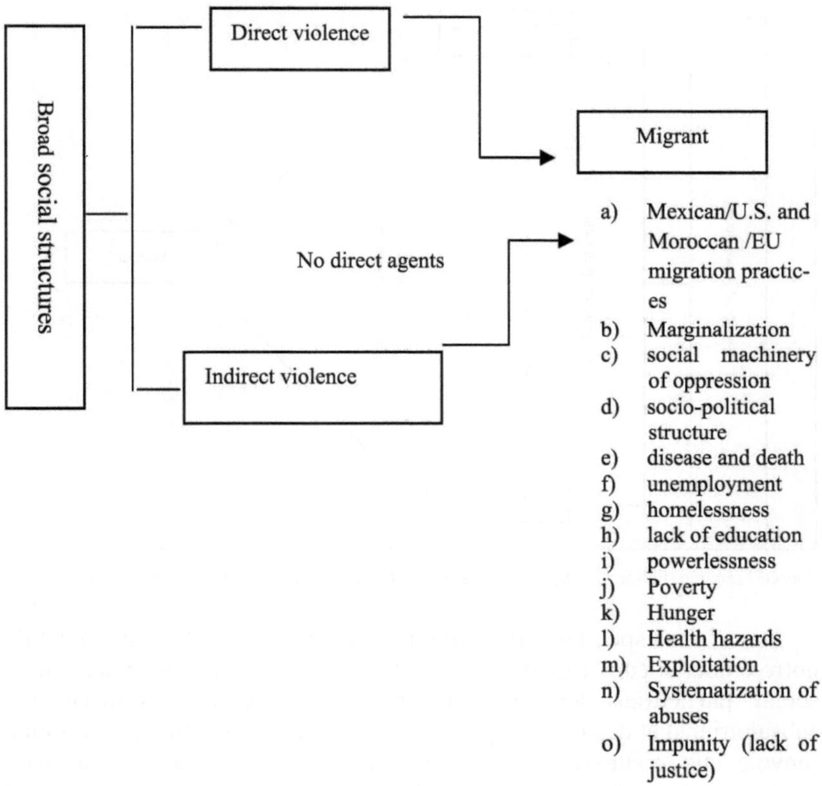

Source : Adaptation de l'auteur inspirée des travaux de Jácome, Felipe (2008). *Trans-Mexican Migration : a Case of Structural Violence*.

Figure 4 : Violence structurelle : impunité et systématisation de la violence

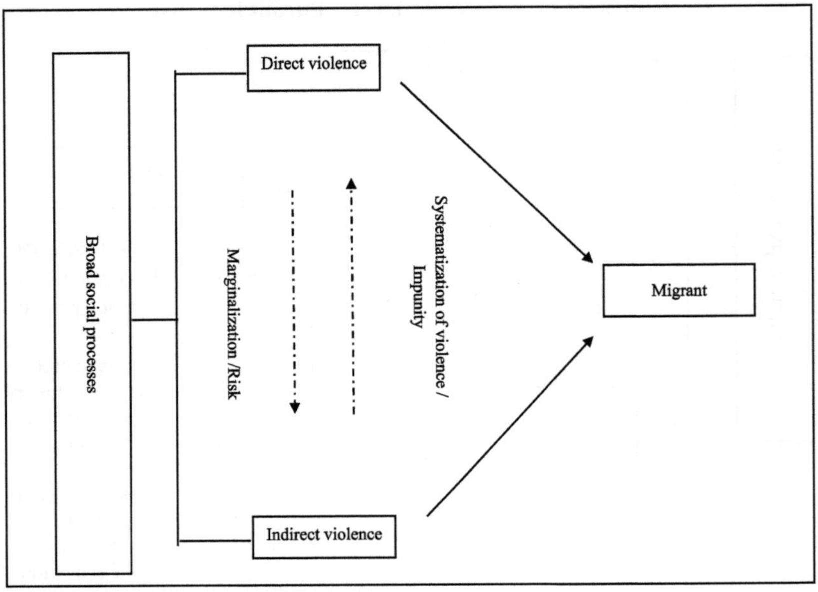

Source : Jácome, Felipe (2008). *Trans-Mexican Migration : a Case of Structural Violence.*

Dans le cas spécifique du Mexique et du Maroc qui font ici l'objet de notre étude, la construction des structures de violence autour du groupe social particulier des migrants irréguliers (centro-sud-américains/ subsahariens) et des cadres spécifiques dans lesquels celle-ci fonctionne renvoie aux contextes historiques relatifs à la violence, l'insécurité économique et politique prévalant dans ces espaces. Il s'agit notamment des violences structurelles qui se situent à l'origine même du phénomène migratoire dans ces espaces et qui incitent les individus à migrer, parmi lesquels les legs issus des guerres, les conflits, la violence et l'instabilité économique et sociale qui traditionnellement prévalent dans ces espaces, etc. Ceci pour dire que la mobilité humaine dans ces espaces prend sa source dans les forces structurelles et les processus historiques majeurs qui nécessairement impactent sur la capacité des personnes à vivre dans leurs communautés d'origine dans la sécurité et la justice. La majorité des recherches s'accordant d'ailleurs pour dire que les conditions structurelles profondes liées à l'instabilité économique et à la violence chronique se situent à la base des motivations des individus à migrer. De ce fait, appréhender la violence subie par les transmigrants au Mexique et au Maroc comme un cas de violence structurelle constitue une excellente approche pour saisir les dynamiques complexes qui caractérisent cette violence, la systématisation de celle-ci, ainsi que l'impunité qui caractérise les forces sociales, politiques et institutionnelles qui la perpétuent. La violence structurelle subie par les migrants irréguliers au Mexique et au Maroc étant par ailleurs perçue comme une résultante des forces transnationales du néolibéralisme, de la démocratisation et de la sécurisation accrue des politiques migratoires implémentées dans les pays d'accueil. Ceci s'applique particulièrement dans le cas des flux migratoires en transit au Maroc et au Mexique dans lesquels se posent les problèmes de l'immigration clandestine, des trafics illicites et de la traite des êtres humains, vis-à-vis desquelles l'UE et les États-Unis adoptent des politiques migratoires restrictives.

III. La violence contre les migrants au Mexique

Le Mexique est un pays d'origine et de transit pour les migrants en direction des États-Unis (USA), mais également un pays de destination pour les migrants en provenance d'Amérique du Sud et centrale. En raison de sa situation géographique (il partage au nord une frontière de 3 152 km avec les États-Unis et au sud une frontière de 956 km avec le Guatemala et de 193 km avec le Belize), le Mexique se présente depuis 2005, comme l'un des plus grands couloirs migratoires au monde, avec près de 400 000 tentatives liées au franchissement illégal des frontières enregistrées chaque année (CNDH, 2013). Pour ces migrants, la traversée du Mexique s'avère généralement être l'une des étapes les plus difficiles, voire la plus dangereuse, de leur parcours migratoire dans le cadre duquel ils seront constamment exposés à la violence et aux abus divers, aussi bien de la part des autorités mexicaines que des criminels de tous bords.

À ce jour, on ne dispose pas du nombre exact des migrants irréguliers (clandestins) présents au Mexique en raison de la nature même des flux. Les statistiques les plus fiables élaborées dans ce sens proviennent majoritairement des données fournies par l'Institut national des migrations du Mexique (INM) et la *US Border Patrol*, fondées pour la plupart sur le solde des arrestations réalisées aux frontières. L'INM et la US Border Patrol s'accordent sur l'augmentation constante ces dernières années, et plus spécifiquement durant la période 2000-2005, du nombre de migrants irréguliers au Mexique et aux abords de la frontière commune avec les États-Unis. Plus de 200 000 cas de détention de migrants irréguliers en provenance d'autres pays que le Mexique recensés en 2005 au Mexique, pour 155 000 arrestations déclarées le long de la frontière sud-ouest des États-Unis (INM, 2005). En 2009, le *Department of Homeland Security* estimait la présence sur le territoire américain (USA) de 6,7 millions ressortissants originaires du Mexique, 530 000 du Salvador, 480 000 du Guatemala et 320 000 du Honduras en situation irrégulière. Le nombre de migrants sans-papiers en provenance de ces pays étant supérieur à celui des migrants en situation régulière (Brick, *et al.*, 2011 ; Hoefner, *et al.*, 2009). En 2010, ces chiffres vont connaître une baisse significative d'après les données fournies par la *US Border Patrol* qui indique notamment l'arrestation de 448 000 migrants à la frontière américano-mexicaine, contre 327 500 enregistrés en décembre 2011. Cela étant très certainement dû aux effets combinés de la crise économique, au renforcement des mesures de contrôle aux frontières et à la multiplication des violences dans cet espace (Nuñez et Heyman, 2007).

Graphique 1 : Solde global des migrants clandestins interpellés par les agents de la U.S. Border, 1995-2012

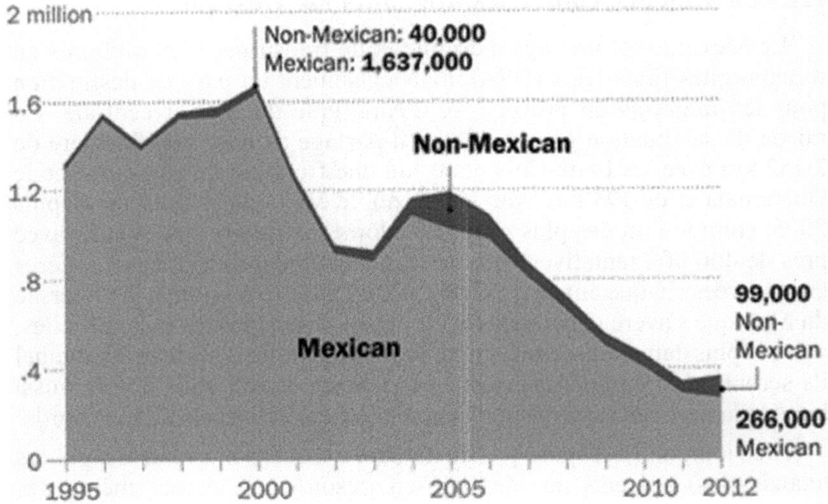

Source : U.S. Customs and Border Patrol and U.S. Department of Homeland Security.

En 2012, l'INM indiquait avoir appréhendé 85 100 migrants, comparativement à 63 072 en 2011. Ces migrants, principalement originaires d'Amérique centrale (Honduras, Guatemala et Salvador), constituent à eux seuls 95 % des personnes détenues par l'INM au Mexique et 89 % des non-Mexicains détenus par les autorités américaines à la frontière des États-Unis / Mexique (INM, 2011). Sur le nombre de migrants expulsés du Mexique en 2012, 45,3 % étaient du Guatemala, tandis que 32,9 % étaient du Honduras, 14,0 % du Salvador, 3,5 % de Cuba et 0,8 % de l'Équateur (INM, 2011). Le graphique ci-dessous réalisé à cet effet par l'INM (2011) nous permet d'avoir une estimation plus précise du nombre de migrants d'Amérique centrale en transit irrégulier au Mexique ainsi que l'évolution de ces flux de 1995-2010.

Graphique 2. Estimations du nombre de migrants d'Amérique centrale en transit irrégulier au Mexique, 1995-2010

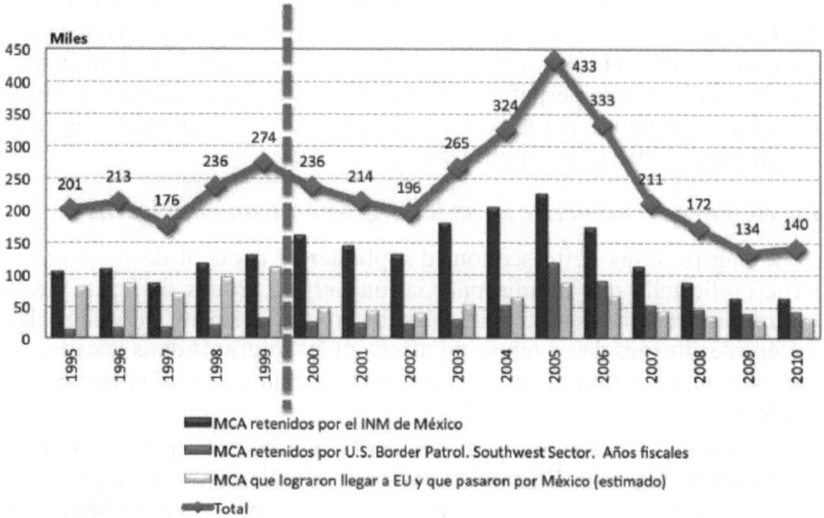

Source : INM, 2011.

Les migrants irréguliers qui entrent au Mexique par Ciudad Hidalgo, Chiapas, à la frontière avec le Guatemala doivent parcourir plus de 3 000 kilomètres pour atteindre les États-Unis. Une fois aux abords de la frontière américano-mexicaine, la traversée des migrants s'opère traditionnellement entre le nord de la Basse-Californie au Mexique et en Californie aux États-Unis. Cependant, le durcissement des politiques migratoires ainsi que le renforcement de la surveillance aux frontières, notamment dans les zones de passage clé, ont entraîné une modification profonde des routes migratoires traditionnelles. Forçant ainsi les migrants clandestins à emprunter des zones isolées ou peu contrôlées, et caractérisées pour la plupart par leur dangerosité, à l'instar du désert de Sasabe situé à la frontière entre l'Arizona et Sonora (COLEF, Anguiano *et al.*, 2007, 2013).

À dater des années 1990, plusieurs chercheurs se sont penchés sur le phénomène de la violence contre les migrants irréguliers ou en situation de transit au Mexique, à l'instar de Casillas (1996, 2001), Castillo (2003), Hagan (2008) Kovic (2008), Mahler (1995), Martinez (2010), Ogren (2007), Spener (2008), Caballero *et al.* (2007) etc. Un sujet jusqu'alors peu étudié et dont certaines lacunes restent toujours d'actualités, en raison notamment du caractère informel des flux et de l'absence d'un registre de données formel indiquant les statistiques sur le nombre de personnes transitant dans cet espace ainsi que sur les événements qui surviennent tout au long du parcours migratoire. Dans l'ensemble, les études réalisées sur le sujet ont été abordées sous trois angles majeurs, dont : (1) la violence relative à la violation des droits des migrants essayant d'entrer clandestinement aux États-Unis ; (2) la violence liée au genre, spécifiquement celle perpétrée à l'encontre des femmes migrantes, et (3) la violence perpétrée par les groupes criminels à l'encontre des migrants originaires d'Amérique centrale et du Sud.

La violence structurelle dans la migration transmexicaine

Il s'agira, dans cette section, d'appliquer le concept de la violence structurelle telle que définie par Galtung *et al.*, au cas de figure de la migration transmexicaine afin d'analyser et de saisir au mieux les différentes formes de violence qui affectent les migrants dans le cadre de leur parcours migratoire vers et à travers le Mexique, dans la perspective finale de se rendre aux États-Unis.

Le Mexique a toujours été un carrefour central de la migration latino-américaine aux États-Unis. Avant les années 1970, la majorité de la population latino migrant vers les États-Unis provenait du Mexique. Toutefois, à dater des années 1970, le Mexique est devenu un pays de transit pour des milliers de Guatémaltèques, Honduriens, et Salvadoriens en route pour les États-Unis. Les États-Unis faisant parallèlement de plus en plus pression sur le gouvernement mexicain pour assurer un meilleur contrôle de sa frontière sud, afin de limiter ces flux (Alba, 2010 ; Alvarez, 1995 ; Castillo, 2006 ; Eschbach, 1999 ; Ogren, 2007). C'est ainsi qu'en 1974, le Mexique va mettre en œuvre, pour la première fois de son histoire, une série de lois restrictives concernant l'immigration. Il s'agit principalement de la *Ley General de Población* (loi générale sur la population – LGP), dont les dispositions porteront sur la pénalisation de la migration clandestine au Mexique (Gonzalez-Murphy, 2011). Dans le contexte relatif aux guerres civiles et à l'instabilité politique prévalant en Amérique Centrale, le Mexique va songer, dans les années 1980, à réviser sa politique à l'égard des sans-papiers en provenance de cette région, notamment à travers l'attribution de l'asile juridique aux réfugiés (Garcia, 2006). Toutefois, la criminalisation des migrants irréguliers restera en vigueur jusqu'au vote unanime du congrès mexicain en 2008, portant sur l'exclusion de loi générale sur la population. En 1993, les pratiques du gouvernement mexicain relatives au contrôle migratoire seront formalisées, avec la création de l'Institut national des migrations (INM) et la mise en place des postes de contrôle le long des principaux axes migratoires, principalement dans le corridor sud-ouest de la côte Pacifique du Chiapas. Par ailleurs, ces mesures juridiques seront accompagnées d'une hausse substantielle du budget consacré à la

« sécurisation » de la frontière sud-mexicaine. Ces mesures auront pour conséquence de renforcer le caractère informel de la migration dans ces espaces, ainsi que la marginalisation des migrants clandestins dans des zones inhospitalières communément appelées au Mexique « *caminos de extravío* » – *chemins perdus*. Sans accès libre aux routes formelles, les migrants irréguliers auront recours à des moyens de transport annexes extrêmement dangereux pour atteindre la frontière des États-Unis, structurés autour d'un ensemble restreint de pistes combinant les routes et les voies ferrées. Créant ainsi un champ fertile à la violence contre les migrants, alimenté par l'impunité caractéristique dont jouissent les agents responsables de celle-ci, ainsi que par la corruption ambiante qui mine la bureaucratie mexicaine. Afin de pallier aux nombreuses critiques suscitées par ces mesures, le gouvernement mexicain va mettre sur pied le Grupo Beta, considéré comme le bras « humanitaire » de l'INM, dont la mission sur le terrain portera paradoxalement sur le contrôle des frontières, l'assistance et la protection aux migrants.

Bien que les changements majeurs dans la politique migratoire mexicaine soient survenus dans les années 1990, d'autres facteurs vont conduire à une militarisation accrue des frontières nord-sud mexicaines, notamment après les attentats terroristes du 11 septembre 2001 aux USA. Sous l'administration du président mexicain Vicente Fox (2000-2006), on assistera à une augmentation exponentielle des fonds portant sur l'implémentation des mesures de contrôle aux frontières, dans l'objectif de limiter les flux migratoires irréguliers en provenance d'Amérique centrale et le trafic de drogue. Le gouvernement mexicain va par ailleurs signer des accords bilatéraux avec le Guatemala, le Belize, le Salvador et le Honduras à cet effet, ainsi qu'avec les États-Unis et le Canada. On peut mentionner, dans ce sens, l'accord pour la création du Groupe de haut niveau pour la sécurité des frontières (GASEF) en 2002, le Partenariat pour la sécurité et la prospérité de l'Amérique du Nord (PNASP) en 2005, etc. C'est toujours dans la même optique que le Mexique va procéder à l'adoption, en 2001, du Plan Sur (Plan Sud) par le biais duquel il entendait non seulement contrôler les flux migratoires irréguliers indexés, mais également lutter contre les narcotrafiquants (Ogren, 2007). Le Plan Sur marquait surtout une nouvelle phase dans la politique migratoire mexicaine, notamment en ce qui concerne le phénomène de l'immigration clandestine, que le gouvernement s'était jusqu'ici forcé d'ignorer. Ceci aura principalement pour effet de criminaliser davantage les sans-papiers à travers le recours justifié à l'usage de la force armée et de la police (Casillas, 2001), la prolifération des centres de détention-rétention pour migrants au Mexique et conduire en conséquence de causes à l'augmentation des taux d'arrestations et de déportations (Hagan, 2008 ; Ogren, 2007). Depuis lors, les projets de militarisation du Plan Sur portant sur la frontière américano-mexicaine ainsi que celle du gouvernement mexicain n'ont cessé d'accroître les mécanismes relatifs au contrôle des flux migratoires et à la détention des migrants comme on peut le voir à travers *l'opération Hold The Line* au Texas, *l'opération Gatekeeper* en Californie et *l'opération Safeguard* en Arizona, etc. Depuis le lancement de *l'opération Gatekeeper* en 1994, la Commission nationale des droits de l'homme au Mexique estime qu'un migrant meurt par jour (CNDH, 2007). En 2008, on recensait 52 camps de rétentions de migrants au Mexique, dont deux en cours de construction et 11 autres en étude (Johnson, 2008). En 2005, l'INM sera intégré au Système national de sécurité du Mexique (*Sistema de Seguridad Nacional de Mexico*). L'accent étant désormais mis sur l'arrestation, la détention et l'expulsion des sans-papiers perçus comme relevant d'une menace pour la sécurité nationale. En février 2011, le gouvernement mexicain proposera une nouvelle loi sur l'immigration (*La Ley de Migración*) finalement votée le 25 mai 2011. Les autorités mexicaines faisant savoir que cette nouvelle loi était destinée à protéger les droits des migrants en situation irrégulière au Mexique, en particulier ceux originaires d'Amérique centrale. Cette loi inclut notamment des dispositions relatives aux soins médicaux et à l'accès à l'éducation pour les migrants sans papiers au Mexique. Cependant, la réalité sera toute autre sur le terrain ou l'impunité prévaut toujours.

Par ailleurs, les États-Unis de leur côté, vont doubler le budget global de leurs dépenses relatives au contrôle des frontières à la suite de l'adoption en 1990, d'une nouvelle Loi sur l'immigration (*The Immigration Act*) ; ouvrant ainsi une ère nouvelle dans la détention des migrants et la dissuasion de l'immigration irrégulière.

L'augmentation des dépenses des États-Unis concernant le contrôle et la sécurisation de sa frontière commune avec le Mexique va connaître une hausse conséquente entre 1986-2010, soit près de 700 millions de dollars en 1986, contre 2,8 milliards de dollars en 2002 et 10,1 milliards de dollars en 2010 (Brick 2011 : 8). Cette dynamique s'observe également sur le terrain à travers la multiplication des agents de patrouille frontalière, soit 9 100 en 2001 contre plus de 20 000 en 2010, la construction d'un « mur » imposant à la frontière, l'équipement d'un matériel de haute technologie et de qualité militaire à l'instar des hélicoptères de combat Blackhawk, des drones-espions sans pilote, des capteurs thermiques, des barrières fictives ou réelles, des caméras et des véhicules blindés. Auxquels s'ajoutent évidemment les agents de patrouille frontalière, les militaires, les gardes nationaux, et les milices de civils armés qui constamment parcourent les zones ciblées à la recherche de clandestins (Brick, 2011 : 8). En 1992, la US Border Patrol comptait 4 026 agents contre 11 106 en 2005, dont près de 90 % étaient affectés aux patrouilles frontalières dans le sud des États-Unis (TRAC Immigration, 2006). Plus spécifiquement, le gouvernement américain va dépenser, en 2010, plus de 17 milliards en vue de sécuriser sa frontière, inclus les 600 millions concédés par le Congrès (Cornelius, 2010). L'investissement massif des États-Unis dans les organes chargés de la sécurisation des frontières, y compris la garde nationale, a généré de nouvelles formes de collaboration entre les deux gouvernements au nom de la sécurité nationale et de la sécurité publique (Curzio, 2007 ; Ramos García, 2005). En nous référant à la typologie de Galtung *et al.*, la militarisation de la frontière américano-mexicaine, ainsi que la sécurisation des politiques migratoires au Mexique et aux États-Unis participeraient ainsi à une forme de violence structurelle imposée aux migrants.

La militarisation de la frontière sud du Mexique va également s'inscrire dans une vague de mesures anti-migratoires et anti-criminelles plus larges, adoptées vis-à-vis des pays d'Amérique centrale et du Sud, à l'instar des initiatives Mérida et CARSI. Entre 2008-2010, le gouvernement américain va allouer des financements estimés à plus de 1,5 milliard de dollars dans le cadre de l'Initiative Mérida adoptée par le Congrès en 2006 sous l'administration Bush, principalement en faveur de la formation et l'achat du matériel d'équipement haute technologie en vue de lutter efficacement contre le crime organisé et le trafic de la drogue au Mexique et en Amérique centrale. Le président Obama, avec l'appui du département d'État américain, ayant par ailleurs sollicité un financement majoré de plus de 310 000 000 $ pour l'exercice de l'année 2011 et 290 millions de dollars pour l'année 2012 dans le cadre du second volet de l'Initiative Mérida (Mérida II). Un faible crédit sera en outre accordé à la dimension des droits de l'homme considérée comme le parent pauvre de cet accord. L'Initiative Mérida, nommée « Plan Mexique » par ses détracteurs, suscitera également quelques interrogations quant à ses accointances avec le très controversé Plan Colombie dans le cadre duquel les États-Unis vont débourser plus de 5,6 milliards de dollars en termes d'aide militaire et la fumigation des champs de coke (Manaut, 2007 ; Velázquez et Schiavon, 2009). La militarisation intensive implémentée par le biais de l'initiative Mérida va indirectement créer des conditions favorables à l'exploitation et au recrutement des migrants irréguliers originaires d'Amérique centrale et du Sud dans les réseaux de trafics illicites et de la drogue, tout en générant de nouvelles formes de violences à leur encontre, particulièrement vis-à-vis des femmes, en témoigne le taux croissant de féminicides observé ces dernières années dans cet espace.

À défaut de mettre définitivement fin aux flux irréguliers dans cet espace, ces mesures sécuritaires vont plutôt alimenter le secteur lié aux trafics illicites, dont les bénéfices seront profitables tant bien aux forces de l'ordre (ex. : agents de police véreux) qu'aux criminels (ex. : trafiquants, passeurs, gangs, cartels), tout en créant des conditions favorables à l'émergence de nouvelles formes de violence contre les migrants irréguliers y compris les populations autochtones résidant dans les régions frontalières (Bacon, 2008 ; Coleman, 2007 ; Zolberg, 2009 ; Andreas, 2012 ; Brotherthon, 2008 ; Nevins, 2010). Ainsi, bien que le passage vers et à travers le Mexique ait été toujours considéré, d'un point de vue historique, comme risqué pour les migrants originaires d'Amérique centrale et du Sud (Menjı'var, 2000), les nouvelles politiques frontalières communément mises en œuvre par les gouvernements mexicain et américain depuis la fin des années 1990 ont fait que ce voyage soit aujourd'hui considéré comme l'une des expériences les plus dangereuses qui soit, et ayant des effets multiples sur la vie des migrants ainsi que celle de leurs familles. Les migrants concernés étant forcés en contrepartie d'adopter de nouvelles stratégies migratoires portant principalement sur le changement des routes et le recours à des trajets de plus en plus isolés ou situés dans des zones inhospitalières afin d'échapper à la vigilance des patrouilles. Le recours monnayé aux passeurs, accroissant de fait leur vulnérabilité face aux criminels d'ordre divers et des gangs associés aux cartels de la drogue. La conséquence à cette situation étant, entre autres, l'augmentation considérable du taux de mortalité des migrants suite à des faits divers relatifs entre autres à la noyade, la déshydratation, les accidents de route, les enlèvements, l'enrôlement de force dans les groupes exerçant des activités illicites, etc. (Cornelius, 2001 ; Eschbach *et al.*, 2001 et 2003). Plusieurs exemples existent dans ce sens. Entre octobre-septembre 2010, l'association No More Deaths indique que les corps de 253 migrants ont été trouvés dans les déserts de l'Arizona (No More Deaths, 2008). En raison de la nature hostile de l'environnement, il est estimé que seuls un cinquième ou tout au plus un dixième des corps sont découverts. En 2006, dix migrants guatémaltèques en route pour les États-Unis ont trouvé la mort à Raudales Malpaso situé à environ 37 km au nord-est de Tuxtla Gutierrez à Chiapas. Les migrants étaient amoncelés à l'arrière d'un camion avec soixante-dix autres Guatémaltèques lorsque celui-ci a heurté un autre camion dans une collision frontale sur un pont. Au moins 16 autres personnes ont été blessées. En 2007, les corps de quinze migrants originaires d'Amérique centrale avaient été récupérés au large de la côte du Pacifique à Oaxaca, près de San Francisco Ixhuatan. D'après le témoignage de quelques survivants, il y aurait eu plus de 20 migrants sur le bateau surchargé qui a chaviré. En 2009, un van transportant des migrants en provenance d'Amérique centrale avait été mitraillé par des agents de police de l'État du Chiapas, et s'était par la suite écrasé contre un arbre. Trois personnes ont été tuées à la suite de cet incident tandis que huit autres ont été grièvement blessés. En 2010, un train de marchandises transportant 300 migrants originaires d'Amérique centrale avait été arrêté par des soldats et des agents de l'immigration mexicains, qui par la suite ont procédé à l'arrestation de 92 personnes. Au cours de ce même voyage, ce train sera une fois de plus attaqué, cette fois par dix hommes armés emportant avec eux 190 personnes dont on ignore jusqu'à ce jour le sort qui leur a été réservé. Les faits décrits ci-dessus ne sont pas isolés et sont monnaie courante au Mexique où ils font régulièrement la une des journaux. Ils illustrent surtout combien le déplacement du point d'origine à la frontière constitue pour de nombreux migrants, à la fois un voyage coûteux qui implique de composer avec les autorités mexicaines corrompues, les bandes criminelles et les trafiquants de drogue, ainsi qu'avec le risque d'être agressé ou blessé au cours du voyage dans le train, d'être arrêtés ou expulsés au Mexique. De plus, la criminalisation croissante des activités liées au franchissement de la frontière, ainsi que la militarisation de celle-ci, a conduit au rapprochement de l'immigration clandestine avec les cartels de la drogue, qui utilisent ces espaces pour faire passer de la marijuana, la cocaïne, l'héroïne et les amphétamines, tout en impliquant les migrants à ces activités. Les liens entre la violence structurelle (indirecte) et personnelle (directe), pour faire référence aux termes de Galtung *et al.*, ainsi que les responsabilités des agents impliqués, étant ici clairement visibles.

Il convient de préciser que le seul instrument juridique dont disposent les migrants irréguliers au Mexique, pour dénoncer les abus ou les violences dont ils sont victimes, se limite au ministère public (Belén, 2008, 11). Toutefois, cette instance est rarement sollicitée par les migrants par crainte de se faire arrêter ou déporter, mais également en raison de la longueur et de l'inefficacité des procédures judiciaire, pour lesquelles les migrants ne disposent pas d'assez de temps pour attendre les sanctions finales. D'un point de vue institutionnel, il n'existe pas d'entités juridiques ou des instruments qui permettent la dénonciation des violations faites aux droits de l'homme par des civils (Bethléem, 2008, 11). Seules les violations des droits de l'homme perpétrées par les autorités gouvernementales peuvent être dénoncées par la Commission nationale des droits de l'homme (CNDH) auprès de l'INM. L'efficacité du CNDH étant cependant remise en cause en raison de son attitude ambiguë, notamment en ce qui concerne sa réticence à dénoncer les actes flagrants de violence commis par les agents de l'immigration mexicaine au cours des « *opérativos* » (ex. : coups et blessures), le manque d'intérêt accordé à ses recommandations par le gouvernement mexicain, les non-lieux réservés à ses plaintes, etc. Les corrélations entre la violence directe et indirecte contre les transmigrants au Mexique s'observent également à travers son caractère systématique. Celle-ci se manifeste dans divers degrés dans les points focaux des principaux axes migratoires entre autres : San Luis de Potosi, Lecheria, Mexico State, Veracruz, Orizaba, Veracruz, Mexico, Coatzacoalos, Chiapas, Ciudad Hidalgo, Tlaxcala, etc. (Casillas, 2006 ; Belen, 2008).

La violence directe ou personnelle

Les faits de violences personnelles perpétrées sur la personne des migrants dans les régions frontalières nord et sud mexicaines ont jusqu'ici été abondamment documentés. Ces faits portent généralement sur l'extorsion, l'abandon en cours de route, le viol, les coups et blessures, les enlèvements, tirs à balles réelles et des accidents de voiture mortels survenus durant le trajet, etc. (Spener, 2009 ; Burnett, 2001 ; Hegstrom, 2001 ; Davis, 2004 ; Winingham y Schiller, 1999). Pour mieux appréhender ce climat de violence, il convient de le situer dans le contexte de l'économie politique relative à la migration clandestine, pour ce qui est spécifiquement du franchissement irrégulier des frontières. Au cours du trajet ou durant le transit des migrants irréguliers vers et à travers le Mexique, on observe un continuum dans la manière dont les individus sont transformés en marchandises et les migrants évalués de façon inégale. On note les différentes façons dont les gens sont perçus, conditionnés et exploités alors qu'ils sont en situation de transit ; selon qu'ils sont destinés au trafic ou à la traite, au travail forcé, à la prostitution, au trafic d'organes ou comme marchandise d'échange. Comparativement aux années précédentes où les migrants étaient dans leur majorité confrontés au vol, aux blessures et mutilations diverses, au viol, à l'extorsion aux abus divers, l'économie politique de la violence apparaît aujourd'hui plus coordonnée et sophistiquée. Le crime organisé, les gangs, les policiers corrompus, l'armée, les autorités en charge de l'immigration et les autochtones mexicains tirant tous profit de la traite, du trafic, des enlèvements et de l'exploitation des migrants clandestins qui vivent ou transitent vers et à travers le Mexique dans l'objectif final de se rendre aux États-Unis. Un processus assimilable à une industrie dont la politique économique tire profit de la mobilité humaine par le biais de la violence structurelle qu'elle génère.

En effet, le renforcement des contrôles aux frontières a affecté à bien des égards les relations entre les migrants et les passeurs aussi bien que leur comportement vis-à-vis de ces derniers. Comme l'a si bien noté Heyman (1999 *et al.*), le renforcement des contrôles augmente les probabilités de rencontre entre les migrants et les coyotes (passeurs, contrebandiers) en vue du franchissement de la frontière. Deuxièmement, ces mesures ont rendu nécessaire le recours aux *coyotes* pour guider les migrants dans les zones isolées ou réputées dangereuses, et dont les conséquences seront visibles à travers l'augmentation des risques liés aux accidents ou à la mort tant bien pour les migrants que les coyotes. Troisièmement, l'intensification de la lutte contre les coyotes ainsi que leur criminalisation par les gouvernements concernés vont contraindre ces derniers à adopter des comportements violents en vue de se protéger ou protéger leurs activités, au détriment de la sécurité des migrants.

Le renforcement accru des contrôles aux frontières pourrait également justifier la démarche des coyotes d'exercer un contrôle financier plus accru sur les migrants en compensation des efforts supplémentaires qu'ils fournissent dans le but de ne pas se faire repérer par les autorités, en suscitant parallèlement la peur vis-à-vis des migrants susceptibles de les dénoncer. De ce fait, les efforts du gouvernement à combattre les *coyotes*, loin de protéger les migrants, auraient plutôt pour effet contraire de les plonger dans une situation de risques accrus.

Il y a peu, un changement notable et alarmant s'est opéré dans la dynamique des violences contre les migrants irréguliers au Mexique, avec la participation de plus en croissante des cartels de drogue dans les activités liées à traite et au trafic des êtres humains, ainsi que dans le kidnapping des migrants à travers le pays. Alors que les violences les plus courantes étaient majoritairement perpétrées par des criminels et les membres des gangs locaux, il y a de cela seulement quelques années, aujourd'hui l'exploitation des migrants ainsi que les abus contre ces derniers sont devenus une partie intégrante des réseaux de trafic de drogue, comme étant l'un des plus grands vecteurs de la criminalité au Mexique et dans la région. Ce qui fait qu'aujourd'hui, la nature des violences perpétrées contre les migrants au Mexique soit en partie le résultat d'une interpénétration accrue des liens entre le prisme sécuritaire des politiques migratoires en vigueur au Mexique, les cartels de drogue, les réseaux de traite et de trafic des êtres humains. Quand ces multiples marchés s'intègrent, ils ouvrent de nouvelles voies propices au profit et à la violence. Dans des rapports plus ou moins récents, le CNDH indiquait que les enlèvements de migrants par les groupes criminels avaient généré à eux seuls près de 25 millions de dollars US de revenus dans la période de six mois comprise entre septembre 2008 et février 2009 (CNDH, 2009, 2011). Ces revenus sont ensuite réinjectés dans les infrastructures du crime organisé dans l'objectif de renforcer leurs capacités sur le terrain, y compris leur emprise sur les migrants, tout en constituant un facteur de risque pour la sécurité régionale. Le groupe de los Zetas illustre parfaitement cette situation. En effet, l'essor et le renforcement de l'emprise de Los Zetas au Mexique ont été un facteur majeur dans l'augmentation du taux d'enlèvements et la commercialisation des migrants sans-papiers. Le groupe de los Zetas a diversifié leurs activités économiques, non seulement pour passer de la drogue, mais aussi pour kidnapper les migrants irréguliers transitant dans cette zone, principalement ceux en provenance d'Amérique centrale et du Sud dont la libération s'effectue généralement en contrepartie d'une rançon et (ou), la contrainte de participer à leurs activités criminelles. À dater de 2007, les Zetas ont commencé à utiliser l'Amérique centrale, principalement le Guatemala, comme base d'importation pour la cocaïne et la méthamphétamine à travers le Mexique (Grupo Savant, 2010). Peu de temps après, les Zetas ont commencé à cibler les migrants originaires d'Amérique centrale avec l'appui d'un certain nombre d'acteurs locaux, inclues les autorités mexicaines. La participation de ces multiples acteurs venant s'ajouter au chaos et à la confusion qui prévalent déjà dans ce contexte.

C'est ainsi que les enlèvements massifs de migrants coordonnés par des actes de violence sont devenus monnaie courante au Mexique, en particulier dans la frontière Sud dans les régions de Chiapas, Oaxaca, Veracruz et Tabasco, qui à elles seules enregistraient 55 % des kidnappings recensés au Mexique entre 2008-2009, contre 67,4 en 2010 (CNDH, 2009, 2011). Ceux-ci sont généralement planifiés par les conducteurs de train, les gangs armés et les passeurs. Les autochtones, les autorités mexicaines et même certains migrants prenant parfois part à ces opérations. Parmi les cas d'enlèvements rapportés par la CNDH entre 2008-2009, 59 % auraient été perpétrés, selon les migrants interrogés, par les passeurs « coyotes » ou « polleros ». Dans ce même rapport, le CNDH va également impliquer, pour la même période, la responsabilité, voire la complicité, des agents du gouvernement mexicain dans 2 % de ces enlèvements, contre à 8,9 % en 2010. À titre d'exemple, l'attaque en décembre 2010 par des agents de l'immigration et forces de police mexicains, avec la complicité du conducteur, d'un train dans l'Isthme de Tehuantepec transportant près de 250 personnes entre Arriaga, Chiapas et Ixtepec, dans l'état d'Oaxaca. Plusieurs migrants ayant été interceptés et forcés de verser un tribut aux intéressés, afin d'être autorisés à continuer le voyage. Ce même train sera stoppé un peu plus loin, cette fois par des hommes lourdement armés qui vont brutaliser et détrousser les voyageurs, et au final, kidnapper 40 personnes. Le gouvernement mexicain ayant même été forcé par la suite, en raison des preuves accablantes, de reconnaître officiellement l'implication de certains de ses agents dans cet événement, dont celle de Ulises Ruiz, alors gouverneur de l'État du Chiapas (*Diario Reforma*, 01/11/07). Les personnes victimes d'enlèvements sont généralement conduites dans des lieux dits « sûrs » où elles sont contraintes de fournir les numéros de téléphone des membres de leur famille établis soit aux États-Unis ou dans leur pays d'origine, que les kidnappeurs contactent par la suite afin de leur exiger une rançon en échange de la vie de leurs proches. Dans un rapport récent, le CNDH (2009) indique qu'entre 2008 et 2009, la rançon exigée variait de 1 500 $ à 5 000 $ avec une moyenne de 2 500 $. Neuf victimes sur 10 (9/10) retenues en captivité déclarant avoir été menacées de leur propre mort ou celle de leurs proches. Plusieurs victimes faisant également mention des brimades diverses, telles que la privation de nourriture, les bastonnades, la séquestration, l'administration forcée des drogues, ou des brûlures, etc. (CNDH, 2009).

Dans le même rapport, le CNDH faisait également état d'un solde global de 9 758 enlèvements de migrants dans la période comprise entre septembre 2008-février 2009, dont la plupart originaires d'Amérique centrale et du Sud. Neuf sur dix (9/10) de ces migrants affirmant que leurs vies étaient réellement en danger (CNDH, 2009 ; Salinas 2009 : 17). En 2010, ce chiffre va connaître une croissance exponentielle, pour se situer désormais à plus de 11 000 incidents d'enlèvements individuels et 214 cas d'enlèvements de masse de migrants recensés sur une échelle de six mois (CNDH, 2011). Ces faits étant devenus de plus en plus communs dans l'Isthme de Tehuantepec, à Oaxaca. Le 24 août 2010, la communauté internationale aura un aperçu brutal de cette réalité intérieure avec la découverte à Tamaulipas au nord-est du Mexique, des corps sans vie de 72 migrants originaires pour la plupart d'Amérique centrale. Les migrants – 58 hommes et 14 femmes, dont l'une était enceinte – ayant été retrouvés les mains liées dans un style d'exécution propre au groupe Zetas qu'ils réservent généralement aux personnes qui refusent de travailler pour eux. En avril 2011, toujours dans la même municipalité, on exhumait les corps de 194 migrants d'une fosse commune. En 2013, Human Rights Watch publiait également un rapport sur les enlèvements au Mexique dans lequel il incriminait la responsabilité des agents de sécurité mexicains dans près de 60 % des cas (Human Rights Watch, 2013). Une réalité qui ne diffère pas réellement de la situation qui prévaut au Maroc.

IV. La violence contre les migrants au Maroc

Pendant longtemps, le Maroc s'est défini comme un pays d'émigration. En 2012, l'IOM estimait à 4,5 millions le nombre de Marocains vivants à l'étranger, soit un dixième de sa population, dont 85 % de ses expatriés établis dans différents pays membres de l'UE, principalement la France, l'Italie et l'Espagne (IOM, 2012). Toutefois, à dater des années 1970-80, le Maroc s'est progressivement imposé comme un pays de transit majeur pour les flux migratoires en provenance d'Afrique sub-saharienne et transitant par le Maghreb, dans l'objectif final de se rendre en Europe (Barros *et al.*, 2000 ; de Haas, 2007 ; Hennebry *et al.*, 2014 ; Khachani, 2010 ; De Haas, 2005a ; Berriane et Aderghal, 2008 ; Bilgili et Weyel, 2009). Le concept de la « migration de transit » sera instrumenté aussi bien par l'UE pour justifier et asseoir son emprise dans la définition des politiques migratoires de son voisinage (Collyer *et al.*, 2012), que par le Maroc comme un levier dans ses négociations avec l'UE et prétexte pour maintenir et consolider sa présence militaire au Sahara occidental (Bredeloup, 2012). En conséquence à cela, une militarisation et un renforcement accrus des contrôles aux frontières, sources d'une plus grande vulnérabilité des migrants (Duvell, 2012). La situation géographique du Maroc, situé aux portes de l'Europe et de l'Espagne (séparés de seulement 14,4 km par le détroit de Gibraltar et les enclaves de Ceuta et de Melilla) n'étant pas également étrangère à cette situation. Le renforcement constant des contrôles aux frontières européennes, ainsi que la « criminalisation » croissante de l'émigration marocaine et des flux migratoires subsahariens en provenance de cet espace vont faire que nombre de ces migrants se retrouvent bloqués pendant de longues périodes au Maroc avant d'être en mesure d'entreprendre ou de poursuivre leur voyage vers l'Europe. Certains étant parfois expulsés du territoire marocain avant d'avoir l'opportunité de franchir les frontières donnant accès à l'Europe. C'est ainsi que le Maroc finira par se définir comme un pays d'immigration-résidence par défaut pour les migrants bloqués (Bredeloup, 2012 ; Collyer, 2010 ; Collyer et de Haas, 2012). Dans la seconde moitié des années 1990, le Maroc fera face à un afflux croissant de migrants et de réfugiés en provenance majoritairement d'Afrique subsaharienne dont la République démocratique du Congo, la République du Congo et la région des Grands Lacs. D'autres flux provenant par ailleurs de la Sierra Leone, du Nigeria et de la Côte d'Ivoire. Une enquête de l'AMERM réalisée en 2007-2008 nous fournit des précisions quant à la répartition par nationalité des migrants subsahariens irréguliers au Maroc, dont : les Nigérians (15,7 %), les Maliens (13,1 %), les Sénégalais (12,8 %), les Congolais (10,4 %), les Ivoiriens (9,2 %), les Guinéens (7,3 %) et les Camerounais (7 %). Les Gambiens (4,6 %), les Ghanéens (4,5 %), les Libériens (3,8 %) et les Sierra-Léonais (3,1 %) apparaissant dans des proportions moindres. La majorité d'entre eux étant basés dans les villes de Rabat, Casablanca, Oujda, Tanger et Laâyoune (AMERM and CISP, 2007 ; AMERM, 2008).

Actuellement, on ne dispose du nombre exact de migrants étrangers en situation irrégulière au Maroc, faute de données fiables sur le sujet et des limites tant quantitatives que qualitatives des études disponibles sur les caractéristiques sociales de ces flux migratoires, y compris les statistiques relatives à la répartition de la population migrante par genre. Toutefois, on recense de nombreuses études relatives à la violence et à l'insécurité qui affectent les migrants irréguliers dans cet espace. Les données recueillies à ce sujet provenant pour la plupart des ONG qui, jusqu'ici, ont publié une série de rapports pertinents, à l'instar de Médecins sans frontières (MSF, 2005, 2008, 2013), Human Rights Watch (2014, 2012) etc. Le caractère constamment mouvant de cette population se situe à la base des principales limites rencontrées dans ce domaine. D'après les dernières estimations fournies le 11 novembre 2013 par les ministères de l'Intérieur et de l'Immigration marocains, leur nombre se situerait entre 25 000 et 40 000 personnes. Des considérations antérieures établies en 2005 et 2008 faisant quant à elles état de 60 000 personnes, dont une majorité d'Européens (74 %) et d'Algériens (19 %) (Fargues, 2007 ; Di Bartolomeo *et al.*, 2009). En ce qui concerne les estimations relatives à la migration irrégulière subsaharienne au Maroc, les données disponibles font état de 10 000 personnes en 2005 (Fargues, 2007), contre

4 500 à 20 000 en 2010 (Khachani, 2010 ; MSF, 2010) et 15 000 en 2013 (ministère de l'Intérieur marocain) ; l'ensemble pris dans le solde global d'une population d'environ 32 millions d'habitants. Ces estimations sont, pour l'essentiel, fondées sur les chiffres des interceptions réalisées aux frontières marocaines, comme l'indique par exemple une étude réalisée par l'AMERM en 2008 et faisant état de l'interception de 9 469 subsahariens en 2006, contre 7 027 en 2007 (AMERM, 2008).

Pour de nombreux migrants subsahariens en partance pour le Maroc, le voyage commence à Agadez au Niger, connu comme étant un carrefour central pour les migrants d'Afrique de l'Ouest projetant de se rendre en Europe. C'est en nombre qu'ils traversent le désert en voiture ou à pied pour atteindre à leur choix, le Maroc, la Libye, la Mauritanie ou l'Algérie (Bensaad, 2005 ; de Haas, 2006, 2008 ; Collyer, 2007). La frontière maroco-algérienne de Oujda – Maghnia constituant pour les migrants, en direction du Maroc, l'ultime obstacle pour accéder au royaume chérifien. C'est à partir de ces deux villes que l'orientation sur le territoire marocain se détermine. Jusqu'en 2005, les migrants subsahariens se rendaient dans la région nord du Maroc (le Rif), le long de la côte méditerranéenne, de Tanger à Nador. Le détroit de Gibraltar et les enclaves espagnoles de Ceuta, en face de Tétouan, et de Melilla face à Nador, ont constitué pendant longtemps, les principaux points de passage des migrants subsahariens. La reconfiguration de ces axes surviendra principalement à la suite de la tragédie de Ceuta et Melilla (2005) et du renforcement des contrôles migratoires dans cet espace. Parallèlement, on va assister au développement des routes de l'axe sud reliant l'archipel des Canaries en pirogue à partir du Sahara occidental, au mépris de leur dangerosité. La traversée en mer à partir de Laâyoune étant également devenue, depuis 2005, un passage fortement prisé par les migrants (Kleist, 2011 ; Bensaad, 2005 ; Lahlou, 2005 ; de Haas, 2006, 2007a, 2007b, 2008).

Aujourd'hui, la violence constitue une réalité quotidienne pour la majorité des migrants subsahariens au Maroc. Durant l'ensemble du processus migratoire, ces derniers sont constamment exposés à une violence criminelle et institutionnelle généralisée, par le biais desquels ils sont sujets à des politiques et des pratiques d'exclusion, de discrimination et de négligence sur la base de leur race, de leur statut, de leur appartenance à un genre spécifique ou de leurs pratiques culturelles et sociales. À cela s'ajoute la violence généralisée commise par les bandes criminelles, y compris par les délinquants et les réseaux de traite et de trafic d'êtres humains, parmi lesquelles des violences sexuelles qui affectent particulièrement les femmes et les mineurs. Associés à cela, les facteurs de vulnérabilité déjà préexistants dans leur pays d'origine. Les auteurs de cette violence agissant généralement sous le couvert d'une impunité totale, car conscients du fait que leurs victimes ne disposent pas d'une protection juridique, policière et médicale réelle en raison de leur statut irrégulier, et par conséquent, qu'ils ne risquent pas de les dénoncer par crainte d'être arrêtés ou exposés à d'autres persécutions.

La violence structurelle dans la migration transsaharienne

Le Maroc fait partie des principaux pays qui ont été ciblés par l'UE dans ses tentatives pour stopper les flux migratoires irréguliers en provenance de l'Afrique. Aujourd'hui, il s'impose comme un partenaire privilégié de l'UE dans le domaine de la coopération migratoire. La problématique portant sur la fermeture des frontières européennes ainsi que sur l'externalisation de leurs politiques migratoires dans les pays tiers a considérablement influencé le changement de politique au Maroc. Dans l'ensemble, les autorités marocaines vont développer une politique faite de fragiles compromis avec l'Union européenne (UE) et l'Espagne, dont on peut voir la matérialisation dans les processus de communautarisation et d'externalisation des politiques migratoires européennes. En passant des traités de Maastricht (1992) et d'Amsterdam (1997), au Conseil de Tampere (1999-2004) et Sommet de Séville (2002), au virage stratégique marqué par le prisme sécuritaire à dater de 2004 à ce jour. Avec entre autres, le programme de La Haye (2005-2009), le pacte européen sur la migration et l'asile (2008), le Programme de Stockholm et le Traité de Lisbonne (2009), ainsi que les différentes communications de la Commission européenne sur les migrations et le développement situées entre 2000-2013, etc.

On retient également dans ce sens les instruments à caractère thématiques tels que les programmes AENEAS et MEDA (partenariat euro-méditerranéen) ainsi que les différents plans indicatifs nationaux (P.I.N.) dans le cadre duquel le Maroc s'est vu fréquemment attribuer des ressources financières afin d'assurer une gestion renforcée de ses frontières et par ailleurs, adapter les instruments législatifs à sa disposition. De décembre 2005 à décembre 2008, le programme AENEAS a bénéficié d'un budget global de 2,5 M €, gérés sous la coordination de la Guardia Civil en tant que partenaire d'exécution au Maroc, en Mauritanie, au Sénégal et au Cap-Vert (Europa.eu). Le programme MEDA constituera, quant à lui, le principal instrument financier du partenariat euro-méditerranéen. Au titre de MEDA II, l'UE a déboursé 67,6 M € entre 2007 et 2010 en faveur du ministère de l'Intérieur marocain pour la « *gestion des contrôles aux frontières* » (Europa.eu). La Commission européenne ayant fourni ces fonds en vue « de renforcer le soutien communautaire au Maroc dans le domaine de la migration, entre autres "*[...] mettre à niveau [....] la politique et les stratégies du gouvernement marocain en matière migratoire [...] améliorer la gestion des flux migratoires et à renforcer la lutte contre l'immigration clandestine*" » (Europa.eu). Ce projet comprenait également, un programme d'urgence destiné à accroître le potentiel des ressources des ministères de l'Intérieur et de l'immigration marocains, dans le domaine de la surveillance et de la mise à niveau des frontières ; et une meilleure efficience de son système judiciaire (Human Rights Watch, 2013). L'instrument de la Politique européenne de voisinage (IEVP) viendra remplacer les programmes MEDA en 2007. Le Plan d'action PEV-Maroc proposant des mesures clés de coopération, dans ce sens, notamment en vue de la « *gestion efficace des flux migratoires et de la signature d'un accord de réadmission avec la Communauté européenne* ». Entre 2007 et 2010, le Maroc a reçu un total de 654 M € dans le cadre de la Politique européenne de voisinage (Commission européenne – PEV 2013), dont près de 130 M € au cours des deux premières années spécifiquement dédiés à la migration (Wunderlich 2012) à travers divers instruments et notamment l'aide au développement. Il serait également intéressant de mentionner dans ce contexte, l'évolution du budget de Frontex dont les financements ont connu une hausse exponentielle, allant de 6 280 202 millions d'euros en 2005 à 87 917 000 millions d'euros en 2010 (Pollack et Slominski, 2009 ; *Journal officiel de l'UE*, 2010).

Le Sommet de Séville (2001) retiendra particulièrement ici notre attention, en ceci qu'il jette les bases juridiques de la politique européenne en matière d'immigration et d'asile, en entérinant officiellement l'introduction systématique d'une « clause migratoire » dans les accords de coopération et d'association de l'UE passés avec les pays tiers (voir Pt. 33 des conclusions de la présidence de Séville). Ce qui n'est rien d'autre que l'expression d'une « conditionnalité migratoire de l'aide au développement », formulée à des fins de gestion des flux migratoires. Le sommet européen de Séville introduit également le principe d'une gestion concertée des flux migratoires à destination de l'Europe, une gestion qui toutefois se situerait en amont de ses frontières c'est-à-dire dans les principaux pays tiers émetteurs de flux (pays de départ et de transit). Il s'agit là d'une délégation partielle du contrôle et de la gestion de ses flux dont elle entend toutefois conserver le monopole, notamment pour ce qui concerne la définition des grandes lignes et du cadre général des décisions élaborées dans ce domaine (Berramdane & Rosetto, 2009 : 54). Par ailleurs, le soutien et l'aide financière européenne deviendront également modulables en fonction de l'efficacité affichée par les pays tiers dans le domaine migratoire, maintenant ainsi une certaine conditionnalité migratoire dans la coopération en matière de développement. Pour ceux des pays tiers qui refuseraient d'apporter leur aimable « contribution » dans la lutte contre le phénomène de l'immigration clandestine, le Conseil de Séville prévoit dans ses conclusions toute une gamme de sanctions plus ou moins indirectes, allant d'une simple mise en berne des relations avec le pays concerné à la suppression définitive de l'aide publique au développement (APE). En combinant donc pression politique et aide financière, l'UE va externaliser une part importante de sa gestion des frontières vers le Maroc, le positionnant parallèlement comme une « nouvelle frontière », ou du moins un « gendarme » de l'Europe, tout en menant invariablement les migrations africaines à être conceptualisée en des termes d'invasion ou de menace. Ceci bien que de nombreuses études aient démontré que la migration africaine constitue une part marginale des mouvements vers l'UE et les pays développés en général (voir les travaux de l'Institut national d'études démographiques – INED ; de Haas, 2007).

Les événements tragiques de Ceuta et Melilla (2005) auront pour effet, de propulser un peu plus les migrations africaines à l'ordre du jour de l'agenda de la politique extérieure européenne. Avec notamment l'adoption en 2005, par le Conseil européen, de l'Approche globale sur la question des migrations, dont les priorités d'actions seront désormais axées sur l'Afrique et la Méditerranée, dans l'objectif majeur de « *réduire les flux de migration clandestine et les pertes en vies humaines, assurer le retour des immigrants illégaux dans de bonnes conditions de sécurité* ». Tout en n'omettant pas d'inclure systématiquement les volets portant sur la gestion des migrations légales et la lutte contre les migrations irrégulières dans les partenariats conclus avec les pays tiers. Les tragiques événements de Ceuta et Melilla donneront également lieu à la première conférence intergouvernementale euro-africaine sur les migrations (Rabat, 2006), menée sur l'initiative de l'Espagne, avec la collaboration du Maroc et de la France. Celle-ci va déboucher sur une déclaration stipulant la nécessité de parvenir à une gestion concertée des flux migratoires en Afrique, ceci à travers la mise en place de projets de développement. Cette conférence sera suivie de près par celle de Tripoli (novembre 2006), d'où il en résultera une déclaration commune visant à inscrire la migration comme un enjeu clé « des stratégies de réduction de la pauvreté » et « de codéveloppement des pays africains » en vue de réduire les flux de migrants et de réfugiés. Il convient également de mentionner les axes du Plan d'action de Rabat à travers lesquels des instruments politiques et financiers multilatéraux et bilatéraux seront également mis en place pour promouvoir la coopération avec l'Afrique sur les questions de migration. Entre autres, la signature des accords bilatéraux de gestion des flux et de réadmission, la signature des conventions de codéveloppement avec attribution de budgets spécifiques, etc.

De manière générale, la contre stratégie du Maroc consistera à instrumentaliser la migration de transit comme un moyen de pression sur l'UE, dans l'objectif évident de parvenir à un accord de libre-échange, de sécuriser les voies légales d'immigration, d'acquérir des droits plus importants pour leurs propres émigrés, et de bénéficier des avantages d'une aide économique consistante et d'une coopération renforcée (L. Martínez, 2009 ; Belguendouz, 2003, 2005, 2010 ; Bensaad 2008 ; Nakache, 2003 ; El Qadim Nora, 2010 ; Gabrielli, 2007, etc.). Foncièrement conscient de la valeur de sa position géographique stratégique unique, le Maroc aura surtout à cœur de se présenter comme un partenaire privilégié et incontournable de l'UE dans l'implémentation de politiques migratoires, mais également comme une opportunité de se différencier à l'échelle régionale. Le cycle de négociations formelles et informelles relatives aux accords de réadmission mettant cependant en relief la marge de manœuvre dont dispose le Maroc dans ses rapports pourtant asymétriques avec l'UE.

Un élément fondamental dans la gestion de l'immigration en transit au Maroc menée sous l'effet de l'influence de l'UE aura a été le développement législatif. C'est le cas de la loi n° 02-03 de 2003 qui jette les bases juridiques relatives à la détention et l'expulsion des migrants clandestins du Maroc, et qui se caractérise par son prisme sécuritaire et le faible crédit accordé aux droits de l'homme ou à la régularisation des migrants. Un fait d'autant plus révélateur qu'elle a été présentée et adoptée au Parlement au même moment que la loi anti-terroriste dite loi 03-03. Une analyse positive de cette loi voudrait que l'on considère son apport dans le processus d'harmonisation de la législation marocaine en matière d'immigration et l'alignement progressif des politiques marocaines sur le principe des droits de l'homme (art. 29). Ces aspects positifs ne sauraient cependant occulter les limites inhérentes à celle-ci. Pas moins de 16 articles de cette loi reprennent fidèlement les textes européens et surtout français en matière migratoire, dont la trame de fond commune reste son prisme sécuritaire. On note particulièrement : l'absence des modalités relatives à la rétention et l'expulsion des migrants (art. 25 à 33) ; l'absence de garanties offertes concernant le respect de leurs droits (limites visibles dans les articles 16-38-65) ; la faible attention accordée au regroupement familial, en violation des engagements internationaux du Maroc contractés lors de la signature de Convention internationale sur la protection des droits des travailleurs migrants et des membres de leur famille en 1993 (ex. : absence d'indications sur les modalités de délivrance des visas aux familles des migrants et du renouvellement de leurs titres de séjour, et des mesures à prendre en cas de décès ou de divorce). On observe également dans cette loi, une tendance générale à la criminalisation du migrant avec un recours prisé et excessif au terme de « *menace à l'ordre public* », conformément au langage utilisé par les pays membres de l'UE. Contribuant ainsi au glissement progressif qui s'est opéré entre les considérations d'ordre migratoire et sécuritaire, particulièrement dans son art. 218.1, qui renvoie au terrorisme. Des dispositions qui ne doivent rien au hasard, en ceci que la référence au terrorisme renvoie nécessairement à traiter les questions migratoires sur la base des principes liés à la défense et à la sécurité nationale et par conséquent de légitimer toute action en la matière, y compris celles qui s'inscrivent en violation des droits de l'homme (Belguendouz, 2003). On note également l'adoption des sanctions pénales sévères à l'encontre des migrants, dont les périodes de détention allant jusqu'à 6 mois (art. 50), ainsi qu'une distinction nette entre « l'entrée et le séjour des étrangers » – parfois irréguliers – et « l'immigration » – toujours « irrégulière » (voir chapitre 8 : titres 1 et 2). Ce qui, selon Delphine Perrin (2008), ne constitue pas moins une manifestation de la volonté du gouvernement marocain « *de réprimer spécifiquement les réseaux de circulation clandestine et d'adresser un message de collaboration aux partenaires européens* ». Elle précise également que cette décision porte à confusion dans la mesure où « l'immigration est entendue en tant qu'entrée sur le territoire, en vue d'un passage [...] et par conséquent, liée à une volonté de sortie vers l'Europe. La circulation de transit étant ainsi, d'office "*présumée*" » (Perrin, 2008). L'immigration étant par la même perçue comme une circulation, détachée du séjour et vouée à la clandestinité. Avec les conséquences que l'on sait aujourd'hui sur les migrants. D'un point de vue idéologie, la loi n° 02-03 marque surtout une nette rupture avec la tradition d'accueil marocaine. L'évolution de la législation migratoire au Maroc sera suivie par le durcissement et le renforcement de leurs pratiques de contrôle, dont les conséquences majeures porteront sur l'augmentation significative des risques et de la vulnérabilité des migrants irréguliers au Maroc. Cette chasse aux migrants va s'opérer aussi bien au Maroc que dans les zones « classiques » de départ des migrants en direction des côtes espagnoles.

Une autre mesure essentielle dans laquelle le Maroc va s'illustrer dans sa collaboration avec l'UE, portera sur l'assistance policière plus ou moins directe dans la surveillance des flux migratoires, principalement en ce qui concerne le contrôle des routes maritimes entre l'Afrique et l'UE. Cette assistance s'illustre notamment par le biais de la formation des forces policières des pays tiers préposées au contrôle des frontières, et également dans l'envoi d'officiers de liaison ou dans la réalisation de patrouilles conjointes le long des routes migratoires informelles les plus empruntées (Gabrielli, 2008). C'est ainsi que le contrôle des côtes va progressivement s'imposer comme un enjeu majeur dans le cadre des requêtes adressées par l'UE et ses États membres au Maroc en matière de contrôle des flux migratoires sur son territoire. Précisons qu'en 1992, le Maroc a conclu un accord de réadmission avec l'Espagne, suivi d'un Protocole d'entente en 2003 et d'un accord de coopération des forces de police en 2012, tous orientés vers la gestion du retour des ressortissants de pays tiers et les ressortissants marocains d'Espagne au Maroc. D'autres États membres de l'UE ont également signé des accords bilatéraux de réadmission avec le Maroc, dont la Belgique (1993), l'Allemagne (1998), la France (1983-1993, 2001), l'Italie (1998), le Portugal (2004), les Pays-Bas (1993), et le Royaume-Uni (2011).

C'est dans cette même logique qu'il convient de situer la signature par le Maroc en juin 2013 du partenariat de mobilité avec l'UE. Un cadre de coopération qui s'inscrit dans la logique de la politique européenne de voisinage et qui reprend ses objectifs en matière migratoire. On constate en effet que l'appel au renforcement et au déploiement d'instruments de lutte contre l'immigration irrégulière tient une place prépondérante dans les propositions retenues dans ce partenariat. Les parties s'engageant par ailleurs à signer, sous peu, un accord de réadmission au Maroc en collaboration avec l'agence FRONTEX. Ce qui ne constitue pas moins une manifestation de plus, de l'externalisation « déguisée » du contrôle migratoire par l'UE. On s'inquiète notamment des risques que cette démarche pose quant au respect des droits des migrants et des réfugiés sur le territoire et aux frontières marocaines, dont la réputation est déjà fortement entachée par les violences et les traitements dégradants perpétrés à l'encontre des migrants. On s'inquiète également du fait que l'engagement général des parties pour l'amélioration des conditions des migrants et des réfugiés reste relativement faible dans cette déclaration, au regard des drames récurrents liés au franchissement des frontières au Maroc. On constate également que les propositions reflètent avant tout les intérêts européens en facilitant l'entrée et le séjour sur le sol, des seules personnes qualifiées, bien que ces derniers s'engagent à promouvoir la mobilité des ressortissants marocains. Une logique donateur/bénéficiaire étant également sous-tendue dans ce partenariat, dont les principes se situent dans le cadre des pratiques relatives à la conditionnalité migratoire de l'aide au développement (marchandage), selon laquelle l'aide économique et les facilitations de visas seraient octroyées en échange du contrôle des « flux migratoires » par le Maroc. Les statistiques proposées par Eurostat à ce sujet sont assez parlantes, ceux-ci situant le Maroc au 7e rang mondial des pays où le plus grand nombre de visas Schengen est délivré : soit 322 094 visas Schengen délivrés en 2012 par les consulats des États Schengen au Maroc et 119 728 permis de séjour délivrés à des Marocains en Europe en 2011 (Eurostat).

Comme on peut le voir, le renforcement des contrôles migratoires exigé par l'UE au Maroc a conduit à des violations diverses ainsi qu'à la vulnérabilité de la protection sociale et juridique des migrants, ainsi qu'à l'usage inapproprié des instruments de coopération technique et financière par les parties en cause et à l'augmentation du pouvoir discrétionnaire et décisionnel des forces de police locales, de l'armée et autres agents affectés au contrôle des frontières. Aggravé par le climat de racisme, de xénophobie et corruption ambiant. Par ailleurs, le renforcement de ces systèmes de contrôle a contribué à l'augmentation des risques auxquels les migrants clandestins sont exposés dans le cadre de leur périple migratoire vers l'Europe, tout comme ils ont contribué à la diversification des routes migratoires empruntées par ceux-ci (les passeurs et les trafiquants étant dans l'obligation de trouver des alternatives à la politique de fermeture des frontières).

La violence directe

En raison de leur statut de « clandestins », la majorité des migrants subsahariens établis au Maroc vivent dans la crainte permanente d'être détenus ou expulsés, ainsi que sous la menace constante de violences, d'abus et d'exploitations divers, de la part d'une série d'acteurs tels que les forces de sécurité marocaines, espagnoles et algériennes, les criminels actifs dans les réseaux de trafic et de traite des êtres humains, les autres migrants et la population civile.

D'après les résultats d'une enquête récente menée par MSF (2013) auprès des migrants subsahariens dans la région de l'Oriental frontalière avec l'Algérie et le territoire espagnol de Melilla (Nador, Oujda, considérée par les experts comme le principal point d'entrée et de sortie des migrants clandestins pour l'Europe), 63 % des personnes interrogées affirment avoir subi des violences au Maroc. Les forces de sécurité marocaines étant dans la plupart des cas indexées (64 %), suivies de loin par les délinquants marocains (21 %) et la Guardia Civil espagnole (7 %). 12 % des cas d'agression impliquant souvent la présence de plusieurs agresseurs. ¾ des victimes interrogées affirmant avoir été victimes de multiples épisodes de violence, situés en moyenne entre 2-5 incidents, contre 5-10 formes de violences pour 14 % et 10 pour 6 % des migrants interrogés. 92 % des victimes affirmant par ailleurs que ces faits de violence avaient été perpétrés de manière consciente et intentionnelle. Dans le même rapport, MSF affirme que de 2010-2012 18 % de ses consultations réalisées sur le terrain sur un échantillon de 2 124 personnes étaient liées aux violences physiques et sexuelles ayant des répercussions psychologiques significatives (MSF, 2013 : 12-13).

Rafles et expulsions

Au Maroc, on assiste régulièrement à des rafles indiscriminées, menées à grandes échelles sur l'ensemble du territoire dans les communautés de migrants subsahariens. Principalement dans les quartiers marginaux des villes telles que Rabat-Salé, Casablanca, Fez et Tanger (MSF, 2013). Au cours de ces raids, les migrants interceptés sont généralement reconduits en masse dans la nuit jusqu'à la frontière algéro-marocaine de Oujda-

Maghnia, indistinctement de leur statut légal (réfugié, demandeurs d'asile, femmes enceintes, mineurs tous confondus) ou de leur situation sanitaire (blessés graves, etc.). Les exposant ainsi à des risques accrus de violences, d'abus, d'exploitation ou d'agressions sexuelles de la part des bandes criminelles qui opèrent le long de cette frontière. Sans oublier la violence exercée par les forces de sécurité algériennes. Les migrants faisant ainsi l'objet d'un « *jeu sinistre de ping-pong* » entre les forces de sécurité établies de l'un comme de l'autre côté de la frontière (MSF, 2013). Dans ce même rapport, MSF (2013 : 15) recense 63 incidents relatifs aux expulsions au compte de l'année 2011, pour un total de 1 300 expulsions enregistrées, parmi lesquelles 38 femmes, dont six enceintes, six mineurs non accompagnés et 24 enfants. Contre 191 incidents d'expulsions enregistrés au cours de l'année 2012 sur un total de 6 000 personnes expulsées. Parmi lesquels 93 femmes, dont 18 enceintes, 45 mineurs, 33 enfants et plus de 500 personnes nécessitant des soins médicaux urgents des suites des actes de violence perpétrés par les forces de sécurité marocaines et/ou espagnoles. Dans leur majorité, les personnes expulsées retournent généralement au Maroc. 68 % des personnes interrogées dans le cadre de l'enquête réalisée par MSF en 2013, déclarant avoir été expulsées depuis leur arrivée au Maroc, dont 80 % plusieurs fois en moyenne (soit 40 % entre 2-5 fois, 25 % entre 5-10 fois et 16 % plus de 10 fois). Précisons que ces données ne sont pas exhaustives et qu'elles devraient probablement être revues à la hausse.

Violences policières des agents marocains et espagnols

Les membres de la société civile, les médias ainsi que nombre d'experts en matière migratoire s'accordent pour dire que les faits de violence directe, dont sont victimes les migrants subsahariens irréguliers au Maroc, sont perpétrés dans leur majorité par les forces de sécurité marocaine, algérienne et espagnole affectées au contrôle des frontières. Dans son rapport publié en 2005, MSF indique que sur 9 350 consultations effectuées sur les migrants entre avril 2003 et mai 2005 dans les villes de Tanger, Nador et Oujda, 2 193 (23,3 %) concernaient des blessures résultant des actes de violence, dont 52 % et 15 % étaient respectivement causées par les forces de sécurité marocaines et espagnoles, 19 % par les bandes de délinquants, 10 % par les réseaux de trafic de personnes et 3 % par certains migrants, contre seulement 1 % jugées comme accidentelles (MSF, 2005 : 6-7). En 2013, ces chiffres étaient estimés à 64 % en ce qui concerne les forces de sécurité marocaines et de 7 % pour la Guardia Civil espagnole, essentiellement dans la région de l'Oriental (MSF, 2013 : 13). Dans la majorité des cas, cette violence survient lors de tentatives de franchissement irrégulier des frontières, lors des raids inopinés des forces de police marocaines dans les camps informels des migrants, des détentions et des opérations de refoulement divers, etc. L'essentiel des blessures survenant dans ce contexte étant la conséquence des violences indirectes occasionnées par les chutes, les tentatives de fuite pour se soustraire aux arrestations des forces de sécurité au cours des rafles, les blessures occasionnées par les barbelés qui recouvrent les multiples clôtures de Ceuta et Melilla, telles que : les fractures, foulures, contusions, plaies, brûlures, commotions cérébrales, amputations plus ou moins graves, blessures sévères, etc. Toutefois, ces faits sont dans leur grande majorité, occasionnés par la violence directe exercée par les forces de sécurité marocaines et espagnoles, parfois accompagnés de civils. Les exemples abondent dans ce sens, souvent soldés par des épisodes malheureux tels que ceux survenus à Ceuta et Melilla dans les mois d'août-octobre 2005 avec la mort de 14 migrants ; le drame du 6 février 2014 à Tarajal, Ceuta, avec le décès de près de 15 migrants au cours d'un raid de la Guardia Civil visant à empêcher l'entrée des migrants clandestins sur le territoire espagnol. Les assauts répétés des barrières de Melilla observés durant l'été 2012 et les violences qui s'en sont suivies. On retiendra globalement dans ce sens : (1) les chasses à l'homme et les raids inopinés des forces de sécurité marocaines dans les camps informels des migrants, à l'instar des rafles tristement célèbres du mont Gourougou menées du 12-14 janvier 2005 avec l'appui de 1 200 membres des forces de sécurité marocaines ; de la forêt de Bel Younech du 9-10 février 2005 avec l'appui de 600 membres de sécurité marocaines ; (2) la déstructuration des campements de fortune, à l'instar des camps informels situés dans les forêts de Moussakine, Sidi Maafa, Mont Gourougou, Selouane, etc. ; (3) la réquisition arbitraire des documents administratifs et des biens personnels des migrants ; (4) les détentions massives et refoulements collectifs de l'Espagne vers le Maroc et du Maroc vers l'Algérie, y compris de mineurs et de femmes enceintes, ainsi que de demandeurs d'asile et réfugiés pourtant « protégés » par la législation marocaine. MSF fait état de 18 319 détentions et reconduites aux frontières enregistrées pour le compte de l'année 2004, dont 2 850 pour la seule zone de Nador ; (5) la dévolution directe aux forces de sécurité marocaines sans aucune procédure légale (ex. : détentions arbitraires, non-assistance juridique et l'absence d'informations données aux migrants concernant les possibilités de demander l'asile, non-traduction des documents soumis à leur signature) ; (6) les traitements inhumains ou dégradants, la non-assistance à personnes en danger, les brutalisassions gratuites et l'usage abusif de la force à l'instar de la traque par des chiens, l'usage des hélicoptères, de la cavalerie et des gaz lacrymogènes, des tirs à balles réelles ou en caoutchouc, l'usage des matraques électriques, voire même l'alliance ad hoc avec les délinquants et les vagabonds locaux (MSF, 2013 ; Human Rights Watch, 2014 ; APDHA, 2014 ; Amnesty International, CIMADE – AFVIC 2004) etc. ; autant d'actions qui provoquent une grave détérioration des conditions de vie des migrants subsahariens déjà si précaires au Maroc. Entraînant parfois des séquelles physiques et psychiques irréversibles.

Traite et trafic des migrants

Depuis quelques années, les réseaux de traite et de trafic d'êtres humains opèrent sur les routes migratoires du Sahel et d'Afrique du Nord. Au cours de cette dernière décennie, leurs activités sont de plus en plus visibles au Maroc, spécifiquement depuis le renforcement des contrôles migratoires aux frontières euromaghrébines mené en vue de limiter l'immigration clandestine en provenance d'Afrique. Forçant ainsi les migrants irréguliers, notamment ceux subsahariens, à avoir recours systématiquement aux services des passeurs et des trafiquants sans lesquels le voyage serait actuellement presque impossible. Dans la pratique courante, ces réseaux criminels ont souvent recours à l'extorsion, aux menaces, à l'intimidation, aux violences physiques et sexuelles, et à la torture afin de s'assurer un maximum de bénéfices économiques et faciliter leurs opérations (MSF, 2013). Le tout dans un climat d'impunité général, convaincus que leurs victimes ne se risqueraient pas à les dénoncer auprès des autorités concernées, sous peine de se faire arrêter, déporter ou brutaliser. Ce qui fait que la violence exercée par ces réseaux reste souvent voilée. Au Maroc, ces réseaux criminels sont coordonnés par une organisation intercommunautaire composée des représentants des différentes communautés migrantes subsahariennes présentes au Maroc (ex. : mafias nigériane et marocaine). On assiste dans certains cas à des kidnappings et à des actes de tortures visant à soutirer de l'argent aux migrants ou à leur famille. De nombreux cas de disparition et de règlement de compte surviennent également dans ce contexte. Le recours aux services des « passeurs » ou « guides » comporte également des risques inhérents, entre autres : le dépouillement des biens personnels des migrants, l'escroquerie, les prestations médiocres (annulation des voyages et non remboursement des frais, abandons en cours de route), les déprédations des groupes criminels. Les actes de violence dirigés contre les femmes et les mineurs étant majoritairement le viol.

V. Conclusion

Aujourd'hui, les routes migratoires au Mexique et au Maroc s'illustrent comme des espaces critiques où la violence ainsi que les inégalités liées au genre ou à la race sont reproduites, et où les relations en rapport avec le genre sont négociées. Une situation qu'il convient également de situer dans le contexte global de la mondialisation et des processus d'intégration économique régionale en cours dans ces espaces. Faute des garanties relatives aux droits de l'homme dans leurs pays d'origine, la vulnérabilité des migrants irréguliers au cours de leur trajet vers et à travers le Mexique et le Maroc s'aggrave en raison de deux facteurs principaux. Le premier étant dû à l'invisibilité ou à la clandestinité dans laquelle les migrants sont forcés de se déplacer : le renforcement des mesures de contrôle aux frontières ainsi que l'application des politiques migratoires restrictives en vigueur au Maroc et Mexique empêchant ces derniers d'avoir respectivement accès aux États-Unis et à l'UE, tout en les forçant à se déplacer de manière clandestine à travers le pays sur des routes de plus en plus dangereuses. Le deuxième facteur étant dû aux contextes de violence prévalant dans ces pays (particulièrement dans le cas du Mexique,) aggravé par la faible attention accordée à cette question par les gouvernements concernés. Bien que les déterminants de majeurs de l'expérience des migrants clandestins au Maroc (subsahariens) et au Mexique (latino-américains) varient considérablement, ils reposent cependant sur des problématiques communes caractérisées par leur statut irrégulier, la discrimination et le racisme, l'exploitation des travailleurs irréguliers, les difficultés d'accès aux services de base, l'exclusion sociale, les conditions de vie précaires, l'insécurité et le risque d'expulsion, etc. (cf. tableau récapitulatif 1 et 2). Bien que les manifestations de la violence directe et indirecte contre les migrants au Maroc et au Mexique parviennent à être identifiées séparément, elles tendent cependant à être une combinaison des deux principaux facteurs qui structurent le quotidien des migrants dans ces espaces. Tout au long de leur périple migratoire, la ligne qui sépare la violence directe et indirecte est flouée par le poids constant des difficultés rencontrées sur le chemin. Les corrélations entre la violence directe et indirecte se matérialisent non seulement dans les événements défavorables affectant les migrants le long du trajet, mais également dans le maintien de l'impunité et la systématisation de la violence. En d'autres termes, la violence indirecte sert de « gardien » à la plupart des formes de violence directe affectant les migrants. Les gouvernements marocain et mexicain s'appuyant même parfois sur la perpétuation de la violence indirecte et l'absence de règle de droit à des fins de contrôle de l'immigration. D'autre part, il est également possible d'observer que la présence de formes directes de violence telles que les viols et les enlèvements le long des itinéraires migratoires, conduit les migrants clandestins dans ces espaces, à subir les effets des violences indirectes à un haut degré. La typologie des violences élaborée par Galtung *et al.* et appliquée aux cas marocain et mexicain nous a permis de constater combien la violence subie par les migrants au cours de leur parcours migratoire dans les espaces étudiés n'est pas statique, mais s'adapte plutôt constamment avec les tendances et les changements migratoires, tout en s'inscrivant en réponse aux évolutions sociopolitiques plus larges. Le parcours des migrants clandestins au Mexique et au Maroc s'inscrit aujourd'hui dans un réseau complexe de violence sociale (violence mettant en évidence les groupes et les réseaux criminels) et de violence institutionnelle (en particulier de l'État) tant dans l'agression directe que dans la criminalisation des migrants. Ce phénomène qui au départ semblait être sporadique et circonstanciel, est rapidement devenu une véritable crise humanitaire qui remet en question non seulement l'essence même des politiques migratoires, mais également, l'ensemble des stratégies gouvernementales qui mettent la sécurité nationale au-dessus des individus et de la sécurité humaine. À l'issue de cette analyse, il apparaît évident que les politiques migratoires implémentées au Mexique et au Maroc, respectivement sous l'influence des États-Unis et de l'UE, sont largement restrictives, et qu'elles peinent toujours à atteindre leurs objectifs. Les contrôles stricts appliqués aux frontières, s'accompagnent de défis majeurs en matière des droits de l'homme, entre autres : les enjeux relatifs à l'asile, la traite et le trafic des êtres humains et la vulnérabilité extrême des migrants en situation irrégulière, et dont la gestion actuelle menace les valeurs et les fondements même de la démocratie libérale. D'où l'intérêt dans ce contexte, de relancer le débat sur le droit éventuel à la mobilité, comme une réponse alternative à la situation au blocage actuel.

Tableau 1 : Violence structurelle dans les pays d'origine des migrants comme le vecteur principal des flux migratoires irréguliers au Maroc et au Mexique

Violence structurelle		
	Mexique (Amérique centrale et du Sud)	**Maroc (Subsahariens)**
Un phénomène résultant des inégalités diverses	**1. Pauvreté et inégalité :** Principaux facteurs à l'origine des flux : la pauvreté, le manque d'opportunités et de services dans leurs communautés d'origine.	**a. La violence et l'instabilité** La propagation des conflits en Afrique subsaharienne (RDC, Congo, Côte d'Ivoire, Mali), comme moteur fondamental de la migration irrégulière au Maroc. Les principaux faits de violence incriminés dans ce contexte étant : la persécution politique, les abus sexuels, les violences conjugales ou communautaires, etc.
	2. Contextes historiques relatifs à la violence : La violence exercée par l'État pendant les périodes de guerre civile en Amérique centrale, dont les populations constituent les flux migratoires majoritaires irréguliers transitant par le Mexique. Avec des conséquences tant au niveau individuel qu'au niveau de la cohésion sociale, telles que la fracture de la cohésion sociale.	**b. Opportunités limitées** il s'agit globalement, du chômage endémique visible dans ces régions, l'absence de moyens d'existence viables, des perspectives d'éducation limitées ou les changements climatiques.
	3. La violence communautaire et urbaine : Taux élevé de criminalité dans les régions d'origine des migrants, la prévalence des pratiques liées à l'extorsion, les homicides, l'insécurité globale. Les auteurs étant non seulement les civils, mais aussi leurs propres gouvernements.	**c. La mobilité et la « culture de la migration »** les réseaux de migrants ainsi que la famille et les proches jouant ici un rôle majeur dans la perpétuation de la migration dans les foyers migratoires.

4. Violence domestique et liée au genre : Principalement en ce qui concerne les violences conjugales	**d. Recherche de meilleures perspectives** Recherche de conditions de vie meilleures liées à l'éducation, l'emploi, les libertés politiques, le besoin de sécurité et autres avantages qui n'existent pas dans les pays de départ. Un corollaire à cela étant l'idéalisation de l'Europe (ou des États-Unis dans le cas du Mexique).
	– Médias et technologies sociales La télévision, Internet et les médias sociaux qui maintiennent la demande. conscience visible des inégalités entre les pays d'origine des migrants et les pays développés (Europe, États-Unis).
	– L'influence de la diaspora Facteur promouvant la migration : Conseil et du soutien financier, symboles des symboles des avantages apparents de la migration.

Source : Adaptation de l'auteur

Tableau 2 : La violence structurelle et directe comme déterminants de majeurs de l'expérience des migrants irréguliers au Maroc et au Mexique

L'expérience des migrants irréguliers au Maroc et au Mexique	
Irrégularité	– Vie dans la clandestinité – Exposition accrue aux abus divers, à l'exploitation et à la violence – Absence d'une protection juridique et policière
La discrimination et le racisme	– stigmatisation de la part des autochtones, de l'État/police et des médias, sur la base des éléments liés à la race, la langue, la religion et de leur statut de résidents irréguliers au Maroc et au Mexique
Emplois insuffisamment payés et relevant de l'exploitation	– Emplois dans les secteurs d'activité les plus dangereux et les moins bien rémunérés, – Perception d'une rémunération nettement inférieure à celle des autochtones – Absence d'une protection sociale recours forcé à la mendicité pour survivre.
Manque d'accès aux services de base	– Difficultés d'accès aux services de base tels que les soins médicaux, l'éducation, l'aide juridique, surtout à l'extérieur des villes. – L'absence d'aide de la part des gouvernements mexicain et marocain relative à leurs besoins
L'exclusion sociale	– hostilité de la part des autochtones mexicains et marocains – marginalisations sociales liées à la nationalité (la langue ou la religion – Maroc), marginalisations au sein même des communautés migrantes (ex : migrants subsahariens au Maroc).

Conditions de vie précaires	– emplois sous-payés – Chômage – coûts élevés de la vie pour les besoins essentiels tels que le logement – pauvreté accrue des migrants – exclusion géographique des migrants dans les zones inhospitalières et les lieux insalubres (forêt, déserts, maisons abandonnées, non-accès à l'eau potable ni électricité, etc.).
L'insécurité et le risque d'expulsion	– Vulnérabilité des migrants face à la violence et au crime – Faiblesse/absence de protection juridique accordée aux migrants – Réticence des migrants à rapporter à la police les faits de violence, d'abus, d'exploitation dont ils sont victimes – Criminalisation des migrants au sein de l'État et de la société

Bibliographie

Alba, F. (2010). México : A crucial crossroads. *Migration Information Source.*

Alexander, T. (1996). *Unravelling global apartheid : An overview of world politics.* Polity press.

Alvarez Jr, R. R. (1995). « The Mexican-US border : The making of an anthropology of borderlands ». *Annual Review of Anthropology*, 447-470.

Amnesty International (2009). *Mexico : New Reports of Human Rights Violations by the Military. Amnesty International.* <https://www.amnesty.org/en/library/asset/AMR41/058/2009/en/e1a94ad6-3df1-4724-a545-f0b93f39af69/amr410582009en.pdf>.

Andreas, P. (2012). *Border games : Policing the US-Mexico divide.* Cornell University Press.

Anguiano, M. E., & Peña, A. T. (2013). « Políticas de seguridad fronteriza y nuevas rutas de movilidad de migrantes mexicanos y guatemaltecos ». *Liminar. Estudios Sociales y Humanísticos*, 5(2), 47-65.

Association marocaine d'études et de recherche en migrations – AMERM and CISP. (2007). *Première enquête sur le phénomène migratoire des subsahariens transitant par le Maroc.* Rabat.

Bacon, D. (2008). *Illegal people : How globalization creates migration and criminalizes immigrants.* Beacon Press.

Barros, L., Lahlou, M., Escoffier, C., Pumares, P., & Ruspini, P. (2000). *L'immigration irrégulière subsaharienne à travers et vers le Maroc.* Programme des migrations internationales, Bureau international du travail.

Belén, P. d. M. (2008). « Humanidad sin fronteras ». *Frontera Con Justicia*, 3.

Belguendouz, A. (2003). *Le Maroc non africain gendarme de l'Europe.* Imprimerie Beni Snassen.

Belguendouz, A. (2005). « Expansion et sous-traitance des logiques d'enfermement de l'Union européenne : L'exemple du Maroc ». *Cultures Et Conflits*, 155-219.

Belguendouz, A. (2005). *UE, Maroc, Afrique migrante : Politique européenne de voisinage, barrage aux sudistes : De Schengen à « Barcelone 10 ».* Abdelkrim Belguendouz.

Bensaad, A. (2005). « Les migrations transsahariennes, une mondialisation par la marge ». *Maghreb-Machrek*, 185, 13-36.

Bensaâd, A. (2008). « Le déplacement des frontières vers le sud ». *Revue Projet*, (1), 50-55.

Berramdane, A., & Rossetto, J. (2009). *La politique européenne d'immigration.* KARTHALA Éditions.

Berriane, M., & Aderghal, M. (2009). *The state of research into international migration from, to and through morocco.*

Bilgili, Ö. & Weyel, S. (2009). « Migration in morocco : History, current trends and future prospects ». *Paper Series : Migration and Development Country Profiles*, 1-62.

Booker, S., & Minter, W. (2001). « Global apartheid ». *The Nation*, 9(11).

Bourdieu, P. (1977). « Sur le pouvoir symbolique ». *Annales*, 405-411.

Bredeloup, S. (2012). « Sahara transit : Times, spaces, people ». *Population, Space and Place*, 18(4), 457-467.

Brick, K., Challinor, A., & Rosenblum, M. R. (2011). *Mexican and Central American immigrants in the United States.* Washington, DC : Migration Policy Institute, June. <Http://www.Migrationpolicy.org/pubs/MexCentAmimmigrants.Pdf>.

Brotherton, D., & Kretsedemas, P. (2008). *Keeping out the other : A critical introduction to immigration enforcement today.* Columbia University Press.

Burnett, J. (2001). « Immigrant smugglers ». *Segment Broadcast on National Public Radio's, Morning Edition*, 22.

Caballero, M., Dreser, A., Leyva, R., Rueda, C., & Bronfman, M. (2002). « Migration, gender and HIV/AIDS in Central America and Mexico ». Paper presented at the *Communication Présentée à La 14e Conférence Internationale Sur Le Sida*, Barcelone, Espagne, 7-12.

Casillas, R. (1996). « Un viaje más allá de la frontera : Los migrantes centroamericanos en México ». *Perfiles Latinoamericanos : Revista De La Facultad Latinoamericana De Ciencias Sociales, Sede México*, (8), 141-171.

Casillas, R. (2000). « Semblanza de la frontera sur de México ». *Varios Autores, Migración : México Entre Sus Dos Fronteras. Foro Migraciones, 2001*, 25-36.

Castillo, M. Á. (2003). « Los desafíos de la emigración centroamericana en el siglo XXI ». *Amérique latine Histoire et Mémoire. Les Cahiers ALHIM. Les Cahiers ALHIM*, (7)

Castillo, M. Á. (2006). « Mexico : Caught between the United States and Central America ». *MIS Selected Readings on Central America*, Migration Policy Institute, Washington, DC. <Http://www.Migrationinformation.org/Feature/display.Cfm>.

Coleman, M. (2007). « Immigration geopolitics beyond the Mexico-US border ». *Antipode*, 39(1), 54-76.

Collyer, M. (2007). « In-Between places : Trans-Saharan transit migrants in morocco and the fragmented journey to Europe ». *Antipode*, 39(4), 668-690.

Collyer, M. (2010). « Stranded migrants and the fragmented journey ». *Journal of Refugee Studies*, 23(3), 273-293.

Collyer, M., Düvell, F., & Haas, H. (2012). « Critical approaches to transit migration ». *Population, Space and Place*, 18(4), 407-414.

Collyer, M., Düvell, F., & Haas, H. (2012). « Critical approaches to transit migration ». *Population, Space and Place*, 18(4), 407-414.

Collyer, M., & Haas, H. (2012). « Developing dynamic categorizations of transit migration ». *Population, Space and Place*, 18(4), 468-481.

Comisión Nacional de los Derechos Humanos (CNDH), coords. (2011). *Informe Especial Sobre Secuestro de Migrantes en México.*

CNDH (2009). *Informe Especial de la Comisión Nacional de los Derechos Humanos sobre los casos de Secuestro en Contra de Migrantes*. México.

CNDH (2007). *Todos saben, nadie sabe : trece años de muerte de migrantes : reporte sobre impunidad y muerte en la frontera sur de Estados Unidos*. México.

CNDH (2013). *Estado debe garantizar atención médica a migrantes. El Universal*, January 19.

Cornelius, W. A. (2001). « Death at the border : Efficacy and unintended consequences of US immigration control policy ». *Population and Development Review*, 27(4), 661-685.

Curzio, L. (2007). *La seguridad nacional de México y la relación con estados unidos.* UNAM.

Davis, V. T. (2004). « Four accused of smuggling migrants ». *San Antonio Express-News*, 24, 1B.

De Haas, H. (2005). *Morocco's migration transition : Trends, determinants and future scenarios.* Global Commission on Internat. Migration.

De Haas, H. (2006). *Trans-saharan migration to north africa and the EU : Historical roots and current trends.* Migration Policy Institute, Migration Information Source.

De Haas, H. (2008). *Irregular migration from West Africa to the Maghreb and the European Union : An overview of recent trends.*

De Haas, H. (2008). « The myth of invasion : The inconvenient realities of African migration to Europe ». *Third World Quarterly*, 29(7), 1305-1322.

Deaths, N. M. (2008). *Crossing the line : Human rights abuses of migrants in short-term custody on the Arizona/Sonora border.* Tucson, United States, NMD [Links].

Di Bartolomeo, A., Fakhoury, T., & Perrin, D. (2009). CARIM-Migration profile : Morocco.

Düvell, F. (2012). « Transit migration : A blurred and politicized concept ». *Population, Space and Place,* 18(4), 415-427.

El Qadim, N. (2010). « La politique migratoire européenne vue du Maroc : Contraintes et opportunités ». *Politique Européenne*, (2), 91-91.

Eschbach, K., Hagan, J., Rodriguez, N., Hernandez-Leon, R., & Bailey, S. (1999). « Death at the border ». *International Migration Review*, 430-454.

Fargues, P. (2007). *Mediterranean migration, 2006-2007 report.* San Domenico Di Fiesole : Robert Schuman Centre for Advanced Studies, European University Institute.

Farmer, P., Bourgois, P., ScheperHughes, N., Fassin, D., Green, L., Heggenhougen, H., Farmer, P. (2004). « An anthropology of structural violence ». *Current Anthropology*, 45(3), 305-325.

Frontières, M. S. (2010). *Sexual Violence and Migration : The Hidden Reality of Sub-Saharan Women Trapped in Morocco En Route to Europe.*

Frontières, M.S. (2013). *Violences, Vulnérabilité et Migration : Bloqués aux Portes de l'Europe, Un Rapport sur les Migrants Subsahariens en Situation Irrégulière au Maroc.* <http://www2.ohchr.org/english/bodies/cmw/docs/ngos/MSF_Morocco18_fr.pdf>.

Gabrielli, L. (2007). Les enjeux de la sécurisation de la question migratoire dans les relations de l'Union européenne avec l'Afrique. *Politique Européenne*, (2), 149-173.

Gabrielli, L. (2008). « Flux et contre-flux entre l'Espagne et le Sénégal. L'externalisation du contrôle des dynamiques migratoires vers l'Afrique de l'Ouest ». *REVUE Asylon(s)*, (3)

Galtung, J. (1969). « Violence, peace, and peace research ». *Journal of Peace Research*, 6(3), 167-191.

Galtung, J. (1985). « Twenty-five years of peace research : Ten challenges and some responses ». *Journal of Peace Research*, 22(2), 141-158.

Galtung, J. (1990). « Cultural violence ». *Journal of Peace Research*, 27(3), 291-305.

Galtung, J., & Fischer, D. (2013). « Cultural violence ». *Johan Galtung* (pp. 41-58) Springer.

Galtung, J., & Höivik, T. (1971). « Structural and direct violence : A note on operationalization ». *Journal of Peace Research*, 8(1), 73-76.

García, M. C. (2006). *Seeking refuge : Central American migration to Mexico, the United States, and Canada.* University of California Press.

González-Murphy, L. V., & Koslowski, R. (2011). *Understanding Mexico's changing immigration laws.* Washington, DC : Mexico Institute, Woodrow Wilson International Center for Scholars.

Grewal, B. S. (2003). *Johan Galtung : Positive and negative peace.* School of Social Science, Auckland University of Technology, 30.

Hagan, J. M. (2008). *Migration miracle.* Harvard University Press.

Hennebry, J., Kopinak, K., Miras, R. M. S., Requena, A. T., & Hondagneu-Sotelo, P. (2014). « From "Khadema" to "Zemegria" : Morocco as a "Migration hub" for the EU ». *Territoriality and migration in the EU neighbourhood*, 65-81, Springer.

Heyman, J. M. (1999). *States and illegal practices.* Berg Oxford.

Hoefer, M., Rytina, N., & Baker, B. C. (2011). « Estimates of the unauthorized immigrant population residing in the United States : January 2010 ». *Population Estimates, Office of Immigration Statistics, Department of Homeland Security*, 4.

Human Rights Watch (2013). *Morocco / Western Sahara : Repression belies reform pledges.* <http://www.hrw.org/news/2013/01/31/moroccowestern-sahara-repressionbelies-reform-pledges>.

Imbusch, P. (2003). The concept of violence. *International handbook of violence research*, 13-39, Springer.

Jácome, F. (2008). *Trans-mexican migration : A case of structural violence.* Center for Latin American Studies, Georgetown University <http://clas.Georgetown.edu/research/papers> (Accessed 31 October 2012).

Johnson, J. (2008). *The forgotten border : Migration & human rights at Mexico's southern border.* Latin America Working Group Education Fund.

Khachani, M. (2010). « Maroc : Migration, marché du travail et développement ». *Document De Travail.* Institut International d'études Sociales, Case Postale, 6.

Khachani, M. (2008). *La migration clandestine au Maroc.*

Kleinman, A. (2000). « The violence's of everyday life ». *Violence and Subjectivity*, 226-241.

Kleist, N. (2011). *Europe fighting irregular migration : Consequences for West African mobility.* DIIS.

Kovic, C. (2008). « Jumping from A moving train : Risk, migration and rights at NAFTA's southern border ». *Practicing Anthropology*, 30(2), 32-36.

Mahler, S. J. (1995). *American dreaming : Immigrant life on the margins.* Princeton University Press.

Manaut, R. B. (2007). La iniciativa Mérida : Desafíos del combate al crimen y el narcotráfico en México (ARI).

Martínez, L. F. (2009). « Les migrations en transit au Maroc. Attitudes et comportement de la société civile face au phénomène ». *L'Année Du Maghreb*, (V), 343-362.

Martínez, Ó. (2010). *Los migrantes que no importan : En el camino con los centroamericanos indocumentados en México.* Icaria.

Menjívar, C., & Abrego, L. (2012). « Legal violence : Immigration law and the lives of Central American Immigrants ». *American Journal of Sociology*, 117(5), 1380-1421.

Mghari, M. (2008). *L'immigration subsaharienne au Maroc.*

Nakache, D. (2003). « Analyse transversale des questions migratoires dans les partenariats économiques nord-sud : Bilan, enjeux et perspectives ». *Communication au « Séminaire EMMA-RINOS, Analyse comparatiste des processus d'intégration régionale Nord-Sud »*, Paris, 26-27.

Nevins, J. (2005). « A beating worse than death : Imagining and contesting violence in the US-Mexico borderlands ». *Ameriquests*, 2(1).

Nevins, J. (2010). *Operation gatekeeper and beyond : The war on « illegals » and the remaking of the US-Mexico boundary.* Routledge.

Núñez, G. G., & Heyman, J. M. (2007). « Entrapment processes and immigrant communities in a time of heightened border vigilance ». *Human Organization*, 66(4), 354-365.

Ogren, C. (2007). « Migration and human rights on the Mexico-Guatemala border ». *International Migration*, 45(4), 203-243.

Penalva, C., & La Parra, D. (2008). « Comunicación de masas y violencia estructural ». *Convergencia*, 15(46), 17-50.

Perrin, D. (2008). « L'étranger rendu visible au Maghreb – la voie ouverte à la transposition des politiques juridiques migratoires européennes ». *REVUE Asylon(s)*, (4).

Pollak, J., & Slominski, P. (2009). « Experimentalist but not accountable governance ? The role of Frontex in managing the EU's external borders ». *West European Politics*, 32(5), 904-924.

Ramos García, J. M. (2005). « Seguridad ciudadana y la seguridad nacional en México : Hacia un marco conceptual ». *Revista Mexicana De Ciencias Políticas y Sociales*, 47(194).

Richmond, A. H., & Valtonen, K. (1994). « Global apartheid : Refugees, racism, and the new world order ». *Refuge : Canada's Journal on Refugees*, 14(6).

Sans Frontières, M. (2005). *Violence et immigration : Rapport sur l'immigration d'origine subsaharienne (ISS) en situation irrégulière au Maroc.* MSF : Barcelona.

Spener, D. (2008). « Global apartheid, coyotaje and the discourse of clandestine migration : Distinctions between personal, structural, and cultural violence ». *Migracion y Desarollo*, 115-140.

Spener, D. (2009). *Clandestine crossings : Migrants and coyotes on the Texas-Mexico border.* Cornell University Press.

Téllez, A., Eugenia, M., & Peña, A. P. T. (2007). « Vigilancia y control en la frontera México-estados unidos : Efectos en las rutas del flujo migratorio internacional ». *Papeles De Población*, 51, 45-75.

Terray, E. (1996). « Réflexions sur la violence symbolique : Autour de Pierre Bourdieu ». *Actuel Marx*, (20), 11-25.

TRAC Immigration. Report date : April 4, 2006. Link. <http://trac.syr.edu/immigration/reports/143>.

Velázquez, R., & Schiavon, J. (2009). *La iniciativa Mérida en el marco de la relación México-estados unidos.* CIDE.

Vogt, W. A. (2013). « Crossing Mexico : Structural violence and the commodification of undocumented Central American migrants ». *American Ethnologist*, 40(4), 764-780.

Zolberg, A. R., & Zolberg, A. R. (2009). *A nation by design : Immigration policy in the fashioning of America.* Harvard University Press.

About the Authors

Vanessa Barolsky is research specialist in the Human Sciences Research Council in Pretoria, South Africa. She worked on a variety of projects on subjects ranging from trust in criminal justice, violence in an international comparative perspective, social cohesion and citizen participation in democratic governance. Before joining the HSRC in November 2006, she worked as an independent researcher – authoring, among other work, the 2003 Annual Aids Review for the University of Pretoria and conducting a major research project for the Centre for the Study of Violence and Reconciliation on Transition and Violence in 2005. Prior to this Dr Barolsky worked as a Director at the Secretariat for Safety and Security as well as serving as a researcher at the South African Truth and Reconciliation Commission where she authored one of the chapters of its final report. (vbarolsky@hsrc.a.za)

Alejandro Agudelo Calle es comunicador social y maestro en Antropología. Actualmente es candidato al doctorado en Ciencias Sociales de la Universidad de Antioquia. Ha sido docente de la Universidad de Antioquia de las maestrías en antropología y comunicación social, de la especialización en investigación social y del pregrado de antropología y de la Fundación Universitaria Luis Amigo en el pregrado de comunicación social. Es investigador del Grupo Recursos Estratégicos, Región y Dinámicas Socioambientales del Instituto de Estudios Regionales de la Universidad de Antioquia. Sus áreas de investigación han sido ciudadanía, participación social, políticas públicas en salud y comunicación pública. Se ha interesado por el campo de la comunicación para la salud, específicamente en el rol que tienen las políticas públicas y programas de promoción de la salud y los medios masivos de comunicación en la formación de ciudadanías en salud. (alejandro.aguadelo@udea.edu.co)

Robert V.H. Dover is Associate Professor of Anthropology at the University of Antioquia in Medellín, Colombia. His works include studies on the indigenous peoples of Canada and Colombia. He co-authored a book entitled (The Right to Health Care: Participation in the Subsidized Health-care Sector). Prof Dover has also recently published on the topic of health and environment in Cape Town, South Africa. (rvhdover@gmail.com)

Paula Andrea Hinestroza has a degree of Magister from the Universidad Pontificia Bolivariana in Medellin. She has written extensively on death, violence, funeral dynamics, and other sociological investigations on the people of Medellin Colombia. Some of her researches appeared in *Papeles de Poblacion* and *Boletín de Antropologia.*

Gloria-Patricia Lopera Mesa is Professor in the EAFIT University's Law School in Colombia. She has a PhD in Law from the University of Castilla – La Mancha in Spain. She has written extensively on various topics where society and law intersects, particularly that of the indigenous in Colombia. Some of her publications are *Principio de proporcionalidad y ley penal: Bases para un modelo de control de constitucionalidad de las leyes penales* (2006) and a chapter in *Race, Ethnicity and National Censuses in Latin American States* (2012).

Émeline Nanga est chercheure en formation doctorale à l'Université du Luxembourg. Son projet de thèse porte sur une étude comparative entre le Mexique et Maroc, concernant le contrôle et la gestion des flux migratoires ; au regard des processus de sécurisation, d'intégration économique et régionale en cours dans ces espaces. (emeline.nanga@uni.lu)

Natasha Elbas Neri is researcher at the Núcleo de Estudos da Cidadania, Conflito e Violência Urbana (NECVU) in the Federal University of Rio de Janeiro (UFRJ) in Brazil. She worked on the topic of narratives of juveniles in Rio de Janeiro in conflict with the law for her PhD at the UFRJ where is also teaches Sociology and Anthropology. (necvu@ufrj.br)

Carlos Mario Perea es exdirector del Instituto de Estudios Políticos y Relaciones Internacionales (IEPRI) de la Universidad Nacional de Colombia y miembro del Concejo Nacional de Paz. Sus temas de trabajo son la cultura política y el conflicto en ciudades de Latinoamérica. Autor de los siguientes escritos: "Limpieza social, una violencia mal nombrada", "Vislumbrar la paz, conflictos en ciudades latinoamericanas", "Porque la sangre es espíritu, cultura política y violencia". (carlosmarioperea@gmail.com)

Claudia Puerta Silva es Antropóloga, Doctora en Antropología Social y Etnología de la Escuela de Altos Estudios en Ciencias Sociales en París. Se desempeña como profesora Asociada del Departamento de Antropología, Facultad de Ciencias Sociales y Humanas. Asimismo, es investigadora Asociada del Grupo Recursos Estratégicos, Región y Dinámicas Socioambientales, Instituto de Estudios Regionales, de la Universidad de Antioquia (UdeA). Sus investigaciones se han focalizado en dos líneas fundamentales: por un lado, ha trabajado sobre conflictos socioambientales ligados a proyectos extractivos y de infraestructura, enfocando su análisis en los procesos de negociación, articulación y resistencia en los cuales se ven involucrados pobladores, empresas privadas y Estado. En este sentido, también se ha interesado desde la economía y la ecología política por entender las dinámicas geopolíticas y económicas que estructuran los regímenes de intervención de estos proyectos en Colombia. (claudia.puerta@udea.edu.co)

Joana Domigues Vargas works at the Núcleo de Estudos em Políticas Públicas de Direitos Humanos in the Federal University of Rio de Janeiro in Brazil. Dr Vargas has a PhD in Sociology and has been working on the topics of security, criminality, juvenile justice and violence. Some of her research appeared in the journals such as *Sociedade e Estado, Educação e Realidade*, and *Dados*. (necvu@ufrj.br)

Regional Integration and Social Cohesion (RISC)

Regional Integration and Social Cohesion (RISC) is an interdisciplinary and multi-lingual (English, French, German, Spanish, Italian and Portuguese) series that examines the human and environmental impacts of regional integration, defined legally/politically (supranationalism), economically (globalisation), socially (identity) and geographically (borders). The series aims to link global analysis of issues associated with social cohesion, such as market shifts, immigration, environmental risk and the deterioration of natural resources, human rights, violence and security, etc., with national and sub-national studies that focus on political decision-making, the non-governmental sector and social participation in public debates, economic transformations, social marginalisation, identity, etc. Thus, the series aims to combine the international, national and sub-national arenas of politics in thematically-oriented research. While proposed case studies are welcome, the works presented in this series from all social science disciplines, are predominantly based on the comparative method. Studies that compare across geographic regions, defined continentally, are of particular interest. In addition to the scientific focus described above, this series aims to connect theoretical analysis of questions related to social cohesion with policy-based research. In doing so, it examines the role of political actors at different levels of regional integration processes and it studies citizen responses to changing opportunity structures in the economic, social and political spheres. Thus, the series attempts to shed light on contemporary shifts in the uses and types of power in policy-making processes. The issue of changes in how policies are being made is linked to: "Who affects policy-making?" and "What impact do policies have in social and economic arenas?"

***Series Editor**: Harlan Koff, Université du Luxembourg*

Published Books

No.15 – Victoria Graham, *Pass or Fail? Assessing the Quality of Democracy in South Africa*, 2015.

No.14 – Pascal Chevalier, *Action locale et développement rural en Europe. Le modèle européen LEADER 2007-2013*, 2014.

No.13 – Lisa Anteby-Yemini, Virginie Baby-Collin, Sylvie Mazzella, Stéphane Mourlane, Cédric Parizot, Céline Regnard & Pierre Sintès (eds.), *Borders, Mobilities and Migrations. Perspectives from the Mediterranean, 19-21st Century*, 2014.

No.12 – Candice Moore, *Regional Integration and Social Cohesion: Perspectives from the Developing World. RISC 2011*, 2013.

No.11 – Claudia Puerta Silva, *Stratégies et politiques de reconnaissance et d'identité. Les Indiens wayuu et le projet minier du Cerrejón en Colombie*, 2013.

No.10 – Carmen Maganda & Olivier Petit (eds.), *Strategic Natural Resource Governance. Contemporary Environmental Perspectives / La gouvernance des ressources naturelles stratégiques. Perspectives contemporaines dans le domaine de l'environnement*, 2012.

No.9 – Peter Gilles, Harlan Koff, Carmen Maganda & Christian Schulz (eds.), *Theorizing Borders Through Analyses of Power Relationships*, 2013.

No.8 – Julia Affolderbach, Travis Du Bry, Olga L. Gonzalez & Constanza Parra (eds.), *Reinforcing Governance. Perspectives on Development, Poverty and Global Crises. RISC 2010*, 2012.

No.7 – Oana Ailenei, *Le rôle de l'économie sociale dans les dynamiques socioéconomiques locales. Des concepts aux initiatives d'innovation sociale à l'échelle des quartiers en Europe et à Roubaix, France*, 2011.

No.6 – Claudia Puerta Silva y Juan Carlos Vélez Rendón (eds.), *Fronteras y reconfiguraciones regionales. RISC 2009*, 2011.

No.5 – Felipe J. Hevia, *Poder y ciudadanía en el combate a la pobreza. El caso de Progresa/Oportunidades de México*, 2011.

No.4 – Carmen Maganda & Harlan Koff (eds.), *Perspectivas comparativas del Liderazgo. RISC 2008 / Comparative Perspectives on Leadership. RISC 2008*, 2009.

No.3 – Harlan Koff (ed.), *Social Cohesion in Europe and the Americas. Power, Time and Space / Cohesión social en Europa y las Américas. Poder, tiempo y espacio*, 2009.

No.2 – Adriana González Gil (ed.), *Lugares, procesos y migrantes. Aspectos de la migración colombiana*, 2009.

No.1 – Harlan Koff, *Fortress Europe or a Europe of Fortresses? The Integration of Migrants in Western Europe*, 2008.

www.peterlang.com